T0199397

Handbook of

Reference Methods for Plant Analysis

Edited by
Yash P. Kalra

Soil and Plant Analysis Council, Inc.

CRC Press
Taylor & Francis Group
Boca Raton London New York

CRC Press is an imprint of the
Taylor & Francis Group, an **informa** business

Published 1998 by CRC Press
Taylor & Francis Group
6000 Broken Sound Parkway NW, Suite 300
Boca Raton, FL 33487-2742

First issued in paperback 2019

No claim to original U.S. Government works

ISBN 13: 978-0-367-44800-4 (pbk)
ISBN 13: 978-1-57444-124-6 (hbk)

**Visit the Taylor & Francis Web site at
http://www.taylorandfrancis.com**

**and the CRC Press Web site at
http://www.crcpress.com**

Library of Congress Cataloging-in-Publication Data

Handbook of reference methods for plant analysis / edited by Yash P. Kalra.
    p. cm.
    "Soil and Plant Analysis Council, Inc."
    Includes bibliographical references and index.
    ISBN 1-57444-124-8 (alk. paper)
    1. Plants—Analysis—Handbooks, manuals, etc. I. Kalra, Yash P.
    II. Soil and Plant Analysis Council. III. Title: Reference methods for
    plant analysis.
QK865.H26   1997
572'.362'072—dc21   97-46558

Library of Congress Card Number 97-46558

FOREWORD

The Soil and Plant Analysis Council, Inc. strives to promote reference methods for soil and plant analysis. In response to this mission, the Council has published since 1974 three editions of a *Handbook on Reference Methods for Soil Analysis*. However, a handbook on reference methods for plant analysis, to the best of my knowledge, is unavailable. In response to this, the *Plant Analysis Handbook* was created.

This *Handbook of Reference Methods for Plant Analysis* is an excellent resource of reference plant methods consolidated into one work. Plant analysis procedures are outlined into easy step-by-step procedures that are laboratory-ready for implementation. Plant laboratory preparation methods such as dry ashing and acid and microwave digestion are discussed in detail, as well as extraction techniques for analysis of readily soluble elements (petiole analysis) and quick test kits for field testing. Other chapters discuss quality assurance/quality control (QA/QC) programs and instrumentation procedures associated with plant analysis procedures.

The intent of this handbook is not to be an exhaustive overview of methods and modifications that exist, but is an attempt to consolidate the *time-tested* methods into one handbook in order to promote standardization of plant analysis procedures.

On behalf of the Soil and Plant Analysis Council, Inc., I want to express our appreciation to Yash Kalra, the authors, the reviewers, and the Council Headquarters staff for the time and effort spent in making this publication possible.

Byron Vaughan
President, 1994–1996

Board of Directors (1995–1996), left to right: J.B. Jones, Secretary-Treasurer, Micro-Macro Publishing; Byron Vaughan, President, Harris Laboratories; Denton Slovacek, HACH Co.; Bob Miller, University of California; Yash Kalra, Canadian Forestry Service; Paul Fixen, Potash & Phosphate Institute; Ann Wolf, Vice President, Pennsylvania State University; Ray Ward, Ward Laboratories; Bob Beck, Vice President, Cenex/Land O'Lakes.

PREFACE

The first edition of the *Handbook on Reference Methods for Soil Analysis* was published by the Soil and Plant Analysis Council in 1974 and then revised in 1980 and 1992. This publication was well received and has proved to be a valuable reference. At the Board of Directors meeting in Seattle, WA in November 1994, it was decided to develop a publication on plant analysis to serve as a complement to the soil analysis handbook. An Editorial Committee was selected and I was asked to serve as Chair of this committee. Much of the planning of the handbook was done by the Editorial Committee during the Board of Directors' meeting in Kansas City, MO in March 1995.

The *Handbook of Reference Methods for Plant Analysis* continues the tradition established when the soil analysis handbook was published by providing laboratories with a standard reference book of analytical methods. This handbook is aimed at a broad audience. It should be a handy reference useful to plant scientists in production agriculture, forestry, horticulture, environmental sciences, and other related disciplines. The methods described are used internationally and have proved to be reliable analytical techniques. The book is designed in a step-by-step format to provide information on state-of-the-art methodology; the procedures are presented in such a way that they can be easily followed and used.

The handbook consists of 27 chapters prepared by 24 authors from Canada and the United States. Contributors are internationally acclaimed experts in their fields. Chapter 25 emphasizes the importance of quality control, with the hope that this will result in the generation of high quality analytical data. Appendix I provides information on the location and selection of appropriate plant material useful for analytical data quality control. This up-to-date compilation enhances the value of Chapter 26 on reference materials for data quality control.

I am indebted to the Council for giving me the opportunity to coordinate this project. I extend my sincere thanks to the members of the Editorial Committee for their cooperation. We are grateful to the authors and the reviewers and all others who contributed directly or indirectly to the publication of this handbook.

Support from the Canadian Forest Service and encouragement from Douglas G. Maynard are gratefully acknowledged. Publishing coordination was done by J. Benton Jones, Jr.

Yash P. Kalra
Editor

SOIL AND PLANT ANALYSIS COUNCIL, INC.

The Soil and Plant Analysis Council, Inc. (formerly the Council on Soil Testing and Plant Analysis) was formed in 1969 in the United States to:

- Promote uniform soil test and plant analysis methods, use, interpretation, and terminology
- Stimulate research on the calibration and use of soil testing and plant analysis
- Provide a forum and an information clearing house for those interested in soil testing and plant analysis
- Bring individuals and groups from industry, public institutions, and independent laboratories together to share information

The officers of the Council consist of a President, President-Elect, and Secretary-Treasurer. The presidency of the Council has been rotated between those who are in the public (usually at a state university) and those in the private sector. The President serves for 2 years. Membership in the Council is open to all. Council membership, including individual, laboratory, and cooperate members, has maintained at approximately 350, of whom 50 are from other than the United States.

Since its formation, the Council has engaged in soil sample exchanges, published proceedings of the Soil-Plant Analysts Workshops, and (co-)sponsored symposia and workshops on soil testing and plant analysis. The Council publishes a quarterly newsletter, *The Soil-Plant Analyst*. In 1974, the Council published the *Handbook on Reference Methods for Soil Testing*, which was revised in 1980. In 1992, a completely revised edition of the handbook was published with the title, *Handbook on Reference Methods for Soil Analysis*. In

1992, the Council published a *Registry of Soil and Plant Analysis Laboratories in the United States and Canada.*

The first Council-sponsored International Symposium on Soil Testing and Plant Analysis was held August 14–19, 1989, in Fresno, CA; the second in Orlando, FL, August 22–27, 1991; the third, August 14–19, 1993 in Olympia, WA; the fourth in Wageningen, The Netherlands, August 5–10, 1995; and the fifth in Bloomington, MN, August 7-11, 1997. The sixth International Symposium will be held in Brisbane, Australia, March 22-26, 1999. All four international symposia were attended by about 200 participants and organized in such a way as to provide maximum discussion and interaction among participants. The proceedings from the four international symposia have been published as special issues of the journal, *Communications in Soil Science and Plant Analysis*, published by Marcel Dekker, Inc., New York, and the proceedings of the fifth symposium will be published soon. Following the first international symposium, it was decided to proceed with these symposia every 2 years, at various locations outside the United States.

In July 1994, the Council initiated a Soil Proficiency Testing Program with 85 participants from both the public and private sectors in the United States and Canada. The program consists of five soil samples sent to participating laboratories in July and again in January. Laboratories are instructed as to what analytes are to be determined and a selection of methods provided. The Soil Proficiency Testing Program was continued for 1995–1996. The Plant Tissue Proficiency Testing program was initiated in July 1995, following the same procedure as that for the Soil Proficiency Testing Program. Five plant tissue samples were sent to participating laboratories in July 1995 and January 1996. Both proficiency testing programs were continued in 1997.

Those interested in all aspects of soil testing and plant analysis are invited to become a member of the Council and Laboratory Members to participate in the Proficiency Testing Programs. Further information can be obtained from the Council Headquarters, Georgia University Station, P.O. Box 2007, Athens, GA 30612-0007, Fax: (706) 613-7573.

Byron Vaughan
Harris Laboratories, Inc.
624 Peach Street
Lincoln, NE 68501
Phone: (402) 476-2811
Fax: (402) 476-7598
E-mail: bvaugl2345@aol.com

THE EDITOR

Yash P. Kalra is an analytical chemist with the Canadian Forest Service, Edmonton, Alberta. He has worked with this department for the last 30 years as head of the laboratory responsible for soil and plant analysis for the Northwest Region. His *Methods Manual for Forest Soil and Plant Analysis*, which he coauthored with Douglas G. Maynard in 1991, is widely used. He coordinated the first soil analysis collaborative study for methods validation by the Association of Official Analytical Chemists (AOAC) INTERNATIONAL. For this study, he received the Methods Committee Associate Referee of the Year Award in 1995. His pioneering work is being used as a model for the upcoming collaborative studies by the Soil Science Society of America and the AOAC INTERNATIONAL.

Yash is a member of the Board of Directors of the Soil and Plant Analysis Council, Inc. and was the 1993 recipient of the *J. Benton Jones, Jr. Award* given by the Council. He is a founder of the Group of Analytical Laboratories (GOAL), a co-founder of the Western Enviro-Agricultural Laboratory Association (WEALA), and has served as President of these two organizations and the Canadian Society of Soil Science. He is Editor, Environmental Sciences, *Journal of Forest Research (Japan),* and chair of the Methods Committee on Environmental Quality, AOAC INTERNATIONAL. Recently, he was appointed a member of the Coordination of Official Methods of Soil Analysis Committee (S889) of the Soil Science Society of America for the 1997–2000 term. He is a Fellow of the Indian Society of Soil Science and the AOAC INTERNATIONAL.

EDITORIAL COMMITTEE

Donald A. Horneck
J. Benton Jones, Jr.
Robert O. Miller
Maurice E. Watson
Ann M. Wolf

CONTRIBUTORS

Tracy M. Blackmer
Monsanto
800 N. Lindbergh
RB4
St. Louis, MO 63167
Phone: (314) 694-2806
Fax: (314) 694-2811
E-mail: TBLAC@monsanto.com

C. Ray Campbell
Agronomic Division
North Carolina Department of Agriculture and Consumer Services
Raleigh, NC 27607-6465
Phone: (919) 733-2655
Fax: (919) 733-2837
E-mail:
ray_campbell@ncdamail.agr.state.nc.us

Dennis D. Francis
USDA-ARS
119 Keim Hall
University of Nebraska, East Campus
Lincoln, NE 68583-0915
Phone: (402) 472-8494
Fax: (402) 472-0516
E-mail: dfrancis@unlinfo.unl.edu

Umesh C. Gupta
Research Centre
Agriculture and Agri-Food Canada
P.O. Box 1210
Charlottetown, PEI, Canada CIA 7M8
Phone: (902) 566-6872
Fax: (902) 566-6821
E-mail: guptau@em.agr.ca

Edward A. Hanlon
Southwest Florida Research & Education Center
Institute of Food and Agricultural Sciences
University of Florida
2686 St. Rd. 29N
Immokalee, FL 34142-9515
Phone: (941) 658-3400
Fax: (941) 658-3469
E-mail: hanlon@gnv.ifas.ufl.edu

Dean Hanson
3017 Ag-Life Sciences
Oregon State University
Corvallis, OR 97331
Phone: (541) 737-5716
Fax: (541) 737-5725
E-mail: hansond@css.orst.edu

Donald A. Horneck
Agri-Check, Inc.
323 6th Street, P.O. Box 1350
Umatilla, OR 97882
Phone: (541) 922-4894
Fax: (541) 922-5496

Milan Ihnat
Pacific Agri-Food Research Centre-
Summerland
Agriculture and Agri-Food Canada
Summerland, BC
Canada VOH 1ZO
Phone: (250) 494-6411
Fax: (250) 494-0755
E-mail: ihnatm@em.agr.ca

Robert A. Isaac
Soil Testing Laboratory
University of Georgia
2400 College Station Road
Athens, GA 30605
Phone: (706) 542-5350
Fax: (706) 369-5734

William C. Johnson, Jr.
Soil Testing Laboratory
University of Georgia
2400 College Station Road
Athens, GA 30605
Phone: (706) 542-5350
Fax: (706) 369-5734

J. Benton Jones, Jr.
Micro-Macro International, Inc.
183 Paradise Blvd., Suite 108
Athens, GA 30607
Phone: (706) 546-0425
Fax: (706) 548-4891

Yash P. Kalra
Northern Forestry Centre
Canadian Forest Service
National Resources Canada
5320 122 Street
Edmonton, AB, Canada T6H 3S5
Phone: (403) 435-7220
Fax: (403) 435-7359
E-mail: ykalra@nofc.forestry.ca

Rigas E. Karamanos
Westco Fertilizer Ltd.
Box 2500
Calgary, AB Canada T2P 2N1
Phone: (403) 279-1120
Fax: (403) 279-1133

C. Grant Kowalenko
Pacific Agri-Food Research Centre
(Agassiz)
Box 1000
Agassiz, BC, Canada V0M 1A0
Phone: (604) 796-2221 Local 216
Fax: (604) 796-0359
E-mail: kowalenkog@em.agr.ca

Liangxue Liu
Lakefield Research Ltd.
185 Concession Street
Lakefield, ON, Canada K0L 2H0
Phone: (705) 652-2000 Extension 2246
Fax: (705) 652-0743

Douglas G. Maynard
Pacific Forestry Centre
Canadian Forest Service
Natural Resources Canada
506 W. Burnside Road
Victoria, BC, Canada V82 1M5
Phone: (250) 363-0722
Fax: (250) 363-0775
E-mail: dmaynard@pfc.forestry.ca

Robert O. Miller
Soil and Crop Sciences
Colorado State University
Fort Collins, CO 80523
Phone: (970) 491-6517
Fax: (970) 491-0564
E-mail: robm846@aol.com

Robert D. Munson
2147 Doswell Avenue
St. Paul, MN 55018-1731
Phone: (612) 644-9716
Fax: (612) 625-2208

C. Owen Plank
Crop and Soil Science Department
Miller Plant Science Building
University of Georgia
Athens, GA 30602-4356
Phone: (706) 542-9072
Fax: (706) 542-7133
E-mail: oplank@uga.cc.uga.edu

James S. Schepers
USDA-ARS
119 Keim Hall
University of Nebraska, East Campus
Lincoln, NE 68583-0915
Phone: (402) 472-1513
Fax: (402) 472-0516
E-mail: jscheper@unlinfo.unl.edu

Denton Slovacek
HACH Company
5600 Lindberg Dr.
Loveland, CO 80539
Phone: (970) 669-3050
Fax: (303) 669-2932
E-mail: dslovace@hach.com

Yoong K. Soon
Agriculture and Agri-Food Canada
P.O. Box 29
Beaverlodge, AB, Canada T0H 0C0
Phone: (403) 354-2212
Fax: (403) 354-8171
E-mail: soony@em.agr.ca

Cornelis (Con) J. Van Laerhoven
Pacific Agri-Food Research Centre
(Agassiz)
Box 1000
Agassiz, BC, Canada V0M 1A0
Phone: (604) 796-2221 Local 247
Fax: (604) 796-0359
E-mail: vanlaerhovenc@em.agr.ca

Maurice E. Watson
Research Extension Analytical
Laboratory
The Ohio State University
1680 Madison Avenue
Wooster, OH 44691-4096
Phone: (330) 263-3760
Fax: (330) 263-3660
E-mail: watson.8@osu.edu

TABLE OF CONTENTS

APPENDICES

PRINCIPLES OF PLANT ANALYSIS

Robert D. Munson

HISTORICAL DEVELOPMENT

The principles of plant analysis have developed since the early 1800s, beginning in Europe (De Saussure, 1804). As various elements were identified as essential for plant growth (Table 1.1), scientists, initially von Liebeg (1840) and subsequently others, began analyzing plants for their content; and a short time later, Weinhold (1862) conceived the idea of using plant analysis as an index of available nutrient element supply. These and other scientists compared plant growth or relative growth or yield with the elemental concentrations contained in the dry matter of the entire plant or plant structures, such as leaves, stems, petioles, fruit or grain, sampled at different times during their development. Goodall and Gregory (1947) reviewed the early research, concluding that much of the work done prior to 1947 could be grouped into one of four categories:

1. Investigations of nutritional disorder made manifest by definite symptoms
2. Interpretation of the results of field trials
3. Development of rapid testing methods for use in advisory work
4. Use of plant analysis as a method of nutritional survey

These categories are still applicable today in terms of research as well as plant analysis utilization in crop production decision-making. Bear (1948) also presented a Western historical perspective of the development of the mineral nutrition of crops that relates to the principles of the plant analysis technique.

1-57444-124-8/98/$0.00+$.50
© 1998 by CRC Press LLC

TABLE 1.1 Discoverer and discoverer of essentiality for the essential elements (Glass, 1989)

Element	Discoverer	Year	Discoverer of essentiality	Year
C	*	*	DeSaussure	1804
H	Cavendish	1766	DeSaussure	1804
O	Priestley	1774	DeSaussure	1804
N	Rutherford	1772	DeSaussure	1804
P	Brand	1772	Vile	1860
S	*	*	vonSachs, Knop	1865
K	Davy	1807	vonSachs, Knop	1860
Ca	Davy	1807	vonSachs, Knop	1860
Mg	Davy	1808	vonSachs, Knop	1860
Fe	*	*	vonSachs, Knop	
Mn	Scheele	1774	McHargue	1922
Cu	*	*	Sommer	1931
			Lipman and MacKinnon	1931
Zn	*	*	Sommer and Lipman	1926
Mo	Hezlm	1782	Arnon and Stout	1939
B	Gay Lussac and Thenard	1808	Sommer and Lipman	1926
Cl	Scheele	1774	Stout	1954

* Element known since ancient times.

Background information on the development of the plant analysis technique is included in a number of reviews and books (Ulrich, 1948; Smith, 1962; Hardy, 1967; Walsh and Beaton, 1973; Reuter and Robinson, 1986; Martin-Prével et al., 1987; Westerman, 1990; Jones et al., 1991; Black, 1993; Mills and Jones, 1996). Marschner (1986) provides historical information on the identification of the essential micronutrients required by higher plants from 1860 to 1954. Glass (1989) has provided a comprehensive list of the discoverer of elements and those establishing their essentiality for plants (Table 1.1). Bergmann (1992) compiled a comprehensive book on nutritional disorders of plants that includes an excellent section on plant analysis.

An element is considered essential if the life cycle of a plant cannot be completed without it, cannot be replaced by any other element, and performs a direct essential function in the plant, criteria that were established by Arnon and Stout (1939).

Elements considered essential for plants include carbon (C), oxygen (O), and hydrogen (H) available from carbon dioxide (CO_2) in the air, through the leaves and water (H_2O) from either rainfall and/or irrigation absorbed by plant roots from the soil; the mineral elements, nitrogen (N), phosphorus (P), potassium (K), calcium (Ca), magnesium (Mg), and sulfur (S), known as the *major elements*; and boron (B), chlorine (Cl), copper (Cu), iron (Fe), manganese (Mn), molybdenum (Mo), and zinc (Zn), known as the *micronutrients*. All these elements are available from the soil and/or are applied as soil amendments, such as lime and fertilizers, and then taken up through the plant roots.

   Sodium (Na), silicon (Si) (Takahashi et al., 1990), and nickel (Ni) (Brown et al., 1987) have been suggested as also being essential to some crop plants, while chromium (Cr), selenium (Se), and vanadium (V) at low concentrations have been found beneficial to some crop plants (Mengel and Kirkby, 1982; Marschner, 1986; Adriano, 1986; Kabata-Pendias and Pendias, 1995; Pais and Jones, 1996), with some of these elements being toxic at higher concentrations. Similar observations have been made by Asher (1991) and Pais (1992). Cobalt (Co) and Ni are essential for the symbiotic fixation of N_2 in soybeans, a process that can dramatically influence the growth of legumes grown on N-deficient soils (Ahmed and Evans, 1960; Dalton et al., 1985).

PLANT CONTENT

The concentration of the essential elements in plants is expressed on a dry-matter basis as either percent or grams per kilogram (g/kg) for the major elements, and either parts per million (ppm) or milligrams per kilogram (mg/kg) for the micronutrients, the units selected depending on the system of use. For grower-farmer use of a plant analysis assay in the United States, the five major elements are normally expressed as percent and the seven micronutrients are expressed in ppm; in the scientific literature, metric or SI units are the units of choice. Equivalent values for the major elements and micronutrients in these various units are as follows (concentration values were selected for illustrative purposes only):

Major elements	*%*	*g/kg*	*cmol(p+)/kg*	*cmol/kg*
Phosphorus (P)	0.32	3.2	—	10
Potassium (K)	1.95	19.5	50	50
Calcium (Ca)	2.00	20.0	25	50
Magnesium (Mg)	0.48	4.8	10	20
Sulfur (S)	0.32	3.2	—	10

Micronutrients	*ppm*	*mg/kg*	*mmol/kg*
Boron (B)	20	20	1.85
Chlorine (Cl)	100	100	2.82
Copper (Cu)	12	12	0.19
Iron (Fe)	111	111	1.98
Manganese (Mn)	55	55	1.00
Molybdenum (Mo)	1	1	0.01
Zinc (Zn)	33	33	0.50

The relative concentration and number of atoms for the essential elements normally found in plant tissue is shown in Table 1.2 and the ranges in concentration in plant leaves associated with insufficiency and sufficiency are given in Table 1.3.

TABLE 1.2 **Average concentrations and relative number of atoms of mineral nutrients in plant dry matter that were considered sufficient for adequate growth (Epstein, 1965)**

Element	Symbol	μmol/g dry wt	mg/kg (ppm)	%	Relative number of atoms
Molybdenum	Mo	0.001	0.1	—	1
Copper	Cu	0.10	6	—	100
Zinc	Zn	0.30	20	—	300
Manganese	Mn	1.0	50	—	1,000
Iron	Fe	2.0	100	—	2,000
Boron	B	2.0	20	—	2,000
Chlorine	Cl	3.0	100	—	3,000
Sulfur	S	30	—	0.1	30,000
Phosphorus	P	60	—	0.2	60,000
Magnesium	Mg	80	—	0.2	80,000
Calcium	Ca	125	—	0.5	125,000
Potassium	K	250	—	1.0	250,000
Nitrogen	N	1,000	—	1.5	1,000,000

Market (1992) has suggested the following average elemental contents expected to occur in plant tissue of an unspecified Reference Plant (Table 1.2a). The elemental concentrations will of course vary with nutrient availability, plant species, growing conditions, and time of sampling.

TABLE 1.2a **Suggested average contents of major, micro and other elements expected to occur in tissue of an unspecified Reference Plant (Market, 1992)**

Major elements	%	Micronutrients	mg/k	Other elements	mg/kg
Carbon (C)	44.5	Boron (B)	40	Aluminum (Al)	80
Hydrogen (H)	6.5	Chlorine (Cl)	20,000	Arsenic (As)	0.1
Oxygen (O)	42.5	Copper (Cu)	10	Barium (Ba)	40
Nitrogen (N)	2.5	Iron (Fe)	150	Cadmium (Cd)	0.05
Phosphorus (P)	0.2	Manganese (Mn)	200	Cobalt (Co)	0.2

Major elements	%	Micronutrients	mg/k	Other elements	mg/kg
Potassium (K)	1.9	Molybdenum (Mo)	0.5	Chromium (Cr)	1.5
Calcium (Ca)	1.0	Zinc (Zn)	50	Lead (Pb)	1.0
Magnesium (Mg)	0.2			Nickel (Ni)	1.5
Sulfur (S)	0.3			Silicon (Si)	10,000
				Sodium (Na)	150
				Strontium (Sr)	50
				Vanadium (V)	0.5

Plant analyses are also used to identify and measure the potentially toxic elements that may be found in plants, such as cadmium (Cd), Cr, mercury (Hg), and lead (Pb), considered disease-producing or toxic in the food chain for humans and animals (Risser and Baker, 1990; Bergmann, 1992; Pais, 1992; Kabata-Pendias and Pendias, 1995; Pais and Jones, 1996). Mengel and Kirkby (1982) have included iodine (I), fluorine (F), and aluminum (Al), in addition to

TABLE 1.3 Approximate concentration or ranges of the major elements and micronutrients in mature leaf tissue generalized as deficient, sufficient or excessive for various plant species

Essential elements	Deficient	Sufficient or normal	Excessive or toxic
		%	
Major elements			
Nitrogen (N)	<2.50	2.50–4.50	>6.00
Phosphorus (P)	<0.15	0.20–0.75	>1.00
Potassium (K)	<1.00	1.50–5.50	>6.00
Calcium (Ca)	<0.50	1.00–4.00	>5.00
Magnesium (Mg)	<0.20	0.25–1.00	>1.50
Sulfur (S)	<0.20	0.25–1.00	>3.00
		ppm	
Micronutrients			
Boron (B)	5–30	10–200	50–200
Chlorine (Cl)	<100	100–500	500–1000
Copper (Cu)	2–5	5–30	20–100
Iron (Fe)	<50	100–500	>500
Manganese (Mn)	15–25	20–300	300–500
Molybdenum (Mo)	0.03–0.15	0.1–2.0	>100
Zinc (Zn)	10–20	27–100	100–400

those mentioned previously as toxic to plants. Even the micronutrients, B, Cl, Cu, Fe, Mn, and Zn, essential at low concentrations (3 to 50 ppm), can decrease growth and become toxic at higher concentrations (20 to 500 ppm).

VISUAL PLANT SYMPTOMS

When an essential element is seriously deficient, photosynthesis or plant metabolic processes are disturbed, and symptoms in terms of the visual appearance of the leaves or decreased growth rate are produced. In addition, crop yield will be seriously decreased unless the deficiency can be corrected. Such visual symptoms are referred to as deficiency symptoms or *hunger signs,* which can be quite specific in terms of a given essential element (McMurtrey, 1948; Sprague, 1964; Grundon, 1987; Bergmann, 1992; Bennett, 1993). Usually a characteristic chlorosis or necrosis of leaf tissue is produced with a stunting of overall growth. The same is true for toxicity symptoms. As the mineral nutrition of a crop is improved, usually both yield and quality are positively affected. For example, grain may have higher mineral, starch, or protein contents, or plumper kernels. Fruits may be juicier or have a greater vitamin content. In some cases, the grade of the grain, fruit, or tuber is improved so that a premium is paid to the producer or marketer. Typical symptoms of essential element deficiency and excess are described in Table 1.4.

TABLE 1.4 Generalized visual leaf and plant nutrient element deficiency and excess symptoms (symptoms may differ significantly between plant species)

Element (symbol)/ status	Visual symptoms
Nitrogen (N)	
Deficiency	Light-green leaf and plant color with the older leaves turning yellow; frequently a V-shaped yellowing and browning of these leaves. Eventually the whole leaf will turn brown and die. Overall plant growth will be slow, the plant itself will be stunted and spindly in appearance, and the plant will mature early with little or no fruit or grain being produced.
Excess	Plants will be dark bluish-green in color and new growth will be succulent. Plants will be easily subjected to disease and insect infestation and to drought stress. The plant will easily lodge. Blossom abortion and lack of fruit set will occur.

TABLE 1.4 Generalized visual leaf and plant nutrient element deficiency and excess symptoms (symptoms may differ significantly between plant species) (continued)

Element (symbol)/ status	Visual symptoms
Ammonium toxicity	Plants fertilized with ammonium-nitrogen (NH_4-N) may exhibit ammonium-toxicity symptoms, with carbohydrate depletion and reduced plant growth. Lesions may occur on plant stems, there may be a downward cupping of the leaves, and a decay of the conductive tissue at the base of the stem with wilting of the plants under moisture stress, due in large part to a cation imbalance or low K. Blossom-end rot of fruit will occur and Mg-deficiency symptoms may also appear. Grain and fruit yield will be reduced.
Phosphorus (P)	
Deficiency	Plant growth will be slow and stunted and the older leaves will have a purple coloration, mainly on the underside of the leaf. Grain and fruit yield will be significantly reduced.
Excess	Phosphorus excess may not have a direct effect on the plant, but the plant may show visual deficiency symptoms of either Zn, Fe, or Mn. High P may also interfere with normal Ca nutrition, with typical Ca-deficiency symptoms occurring.
Potassium (K)	
Deficiency	On the older leaves, the edges will look burned, a symptom known as *scorch*. Plants lodge easily, are sensitive to disease infestation, will be stunted, and senesce prematurely. Fruit and seed production will be impaired and of poor quality. Poor post-harvest quality has been frequently related to inadequate K, although visual plant symptoms may not be evident.
Excess	Plants will exhibit typical Mg, and possibly Ca, deficiency symptoms due to a cation imbalance.
Calcium (Ca)	
Deficiency	The growing tips of roots and leaves will turn brown and die. The edges of the leaves will look ragged as the margins of emerging leaves tend to stick together. If the deficiency is severe, leaves may never fully emerge; and for corn, emerging leaves will stick together, giving a *ladder-like* appearance to the plant. Fruit quality will be affected with the occurrence of blossom-end rot on fruits. Lack of adequate Ca may result in the decay of the lower stem conductive tissue, with plants easily wilting when a high evaporative demand exists.

TABLE 1.4   Generalized visual leaf and plant nutrient element deficiency and excess symptoms (symptoms may differ significantly between plant species) (continued)

Element (symbol)/ status	Visual symptoms
Calcium (continued) *Excess*	Plants may exhibit typical Mg deficiency symptoms, and when in high excess, K deficiency may also occur.
Magnesium (Mg) *Deficiency*	Older leaves will be yellow in color with interveinal chlorosis (yellowing between the veins) symptoms with some reddish purpling of these affected leaves. Within the chlorotic stripe, sections of the stripped area will die, giving a beaded-streaking appearance. Plant growth will be slow and some plants will become sensitive to disease infestation. Depending on other conditions, blossom-end rot of fruit may be induced.
Excess	Of rare occurrence, but when occurring will result in a cation imbalance, the plant showing signs of either a Ca and/or K deficiency.
Sulfur (S) *Deficiency*	An overall light-green color of the entire plant with the older leaves grading to yellow in color as the deficiency intensifies. Plants will be slow growing, stunted with delayed maturity, and low yielding.
Excess	A premature senescence of leaves may occur.
Boron (B) *Deficiency*	Abnormal development of the growing points (meristematic tissue) with the apical growing points eventually becoming stunted and dying. Flowers and fruits will abort. For some grain seed and fruit crops, yield and quality are significantly reduced. Pollination is reduced by B deficiency. Internodes may not elongate, giving a compressed appearance to the plant.
Excess	Leaf tips and margins will turn brown and die. When severe, the whole plant will be stunted and may die since elevated B levels can be toxic to many plants.
Chlorine (Cl) *Deficiency*	Younger leaves will be chlorotic and plants will easily wilt. For wheat (as well as other small grains), disease will infest the plant when Cl is deficient.

TABLE 1.4 Generalized visual leaf and plant nutrient element deficiency and excess symptoms (symptoms may differ significantly between plant species) (continued)

Element (symbol)/ status	Visual symptoms
Excess	Premature yellowing of the lower leaves with burning of the leaf margins and tips. Plants will easily wilt, and leaf abscission will occur, primarily for woody plants.
Copper (Cu) *Deficiency*	Plant growth will be slow and plants stunted with distortion of the young leaves and death of the growing point. The plant will appear limp and the margins of the older lower leaves may become necrotic.
Excess	Fe deficiency may be induced with very slow growth. Roots may be stunted and the root tips may die.
Iron (Fe) *Deficiency*	Interveinal chlorosis will occur on the emerging and young leaves with eventual bleaching of the new growth. When severe, the entire plant may be light-green in color. For some plants, growth will be slowed.
Excess	Not of common occurrence, but when occurring there is a bronzing of leaves, and tiny brown spots, a typical symptom, frequently occurring with rice.
Manganese (Mn) *Deficiency*	Interveinal chlorosis of young leaves while the leaves and plants remain generally green in color (the green color making Mn-deficiency symptoms distinct from Fe deficiency). When severe, white streaks will appear on the leaves of some plants, and plant growth will be slow. The deficiency will affect the appearance and quality of forage legumes, particularly alfalfa.
Excess	Older leaves will show brown spots surrounded by a chlorotic zone and circle. Small black specks will appear on the stems and tree fruits, a condition frequently referred to as *measles*.
Molybdenum (Mo) *Deficiency*	General symptoms frequently appear similar to that for N. Older and middle leaves become chlorotic first and, in some instances, leaf margins are rolled, new growth is malformed, and flower formation restricted.

TABLE 1.4   Generalized visual leaf and plant nutrient element deficiency and excess symptoms (symptoms may differ significantly between plant species) (continued)

Element (symbol)/ status	Visual symptoms
Molybdenum (continued)	
Excess	Not of common occurrence.
Zinc (Zn)	
Deficiency	Upper leaves will show interveinal chlorosis with an eventual whitening on either side of the mid-rib of the affected leaves. Leaves may be very small and distorted with a rosette form. Internodes will be short and plants will be stunted.
Excess	Fe deficiency will develop. When toxicity is severe, plants will be severely stunted and will eventually die.

SAMPLING AND METHODS OF ANALYSIS

Before a plant analysis result could be used effectively, research had to be conducted to relate elemental concentrations in the plant-to-growth response or yield. Also, it is necessary to determine what part or portion of the crop or forage plant, vine, or tree should be sampled and the time for sampling. Sample timing usually involves some point in the physiological age or stage of development of the crop, the selection being influenced as to whether the plant is an annual or perennial. Specific sampling procedures are given in the next chapter.

Plant analyses are usually carried out on prepared samples in a laboratory under controlled conditions (Jones and Case, 1990). Testing of forage samples may be conducted with near infrared spectrometry (NIR) in mobile laboratories. Plant tissue tests or quick tests (see Chapter 19) can be conducted in the field on living plant tissue, plant sap, or fluid (Krantz et al., 1948; Morgan and Wickstrom, 1956; Ohlrogge, 1962; Wickstrom, 1963, 1967; Syltie et al., 1972; Jones, 1994a). The Potash and Phosphate Institute (PPI)* has a slide set on the field use and interpretation of tissue tests that can be helpful for those wishing to see how these tests can be used to augment plant analysis. Jones (1994b,c) has videos on plant analysis and tissue testing. Quick tests, such as nitrate-N (NO_3-N), can be determined on plants in the field, and samples taken for

---

* 655 Engineering Drive, Suite 110, Norcross, GA 30092-2843.

laboratory analyses to verify the results. The use of non-destructive techniques, such as chlorophyll meters (see Chapter 20) and infrared photography, are techniques that will be used increasingly to monitor the nutritional status of a crop or plant during its life cycle.

Sample preparation and analysis procedures are covered in detail in other chapters in this handbook.

IMPORTANT ASPECTS OF PLANT ANALYSIS

There are a number of important things to keep in mind when using the plant analysis technique. Some of these have been discussed by Ulrich and Hills (1967) and Munson and Nelson (1973, 1990). Critical nutrient element levels have to be established by restricted growth comparisons and correlation studies for each element and for each crop under a variety of conditions. Ulrich (1948) and Smith (1962) reviewed the early developments and uses of plant analyses. Steenbjerg (1951) discussed different yield curves as related to plant nutrient concentrations. Munson and Nelson (1973, 1990) discussed the interrelationships of yield curves, nutrient supplies, and nutrient concentrations, and also related relative yield to the deficient zone, transition zone or critical range, adequate zone or sufficiency range (optimum concentration occurs at or near maximum or optimum economic yield), and excess zone (occurs at nutrient concentrations for which relative or actual yields begin to decrease). It should be noted that for nearly every element above the optimum concentration, there is an adequate zone or sufficiency range, which indicates that most elements can be taken up at levels greater than needed for *optimum* yield before yields are decreased due to an excess or an imbalance with other elements occurs. Some refer to this as the *luxury range* (Bergmann, 1992), inferring luxury consumption, which appears to occur for most elements. Examples of the different types of relationships that can be used in studying plant analysis results are shown in Figures 1.1 to 1.3.

Figure 1.1 is the mineral nutrition concept given by Macy (1936) relating yield and nutrient concentration in the crop to a single nutrient supply. Figure 1.2 relates the relative growth or percent of maximum yield to the nutrient concentration of an element to various nutritional zones. Some may accept a nutrient concentration that produces 90% of maximum yield in the transition zone as the *critical level*. I would prefer to accept nutrient concentrations that produce the maximum or optimum yield as the critical or optimum levels. The adequate zone would correspond to the sufficiency range. Figure 1.3 provides another way of perceiving relationships from the standpoint of a grower or farmer.

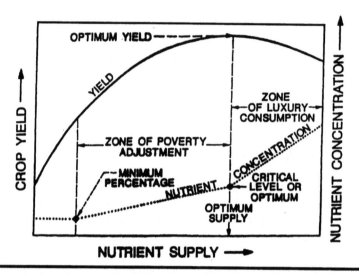

FIGURE 1.1 Schematic of the relationship of crop yield and nutrient concentration as influenced by the supply of a single nutrient (Macy, 1936; Munson and Nelson, 1990). Used with permission of the Soil Science Society of America, Madison, WI.

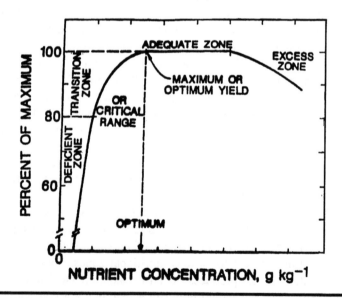

FIGURE 1.2 Schematic of the relationship between the percentage of maximum or relative yield, deficient, transition or critical range, adequate, and excess zones as influenced by increasing concentrations of single nutrient (Ulrich and Hills, 1990; Dow and Roberts, 1982; Munson and Nelson, 1990). Used with permission of the Soil Science Society of America, Madison, WI.

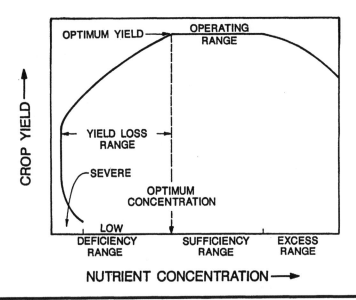

FIGURE 1.3 Schematic of crop yield or growth as related to increasing nutrient concentrations through the deficiency, sufficiency, and excess ranges (Chapman, 1966; Munson and Nelson, 1990). Used with permission of the Soil Science Society of America, Madison, WI.

It should be remembered that yield curves, whether curvilinear or linear-plateau, are always estimated with some degree of error and that confidence limits must be determined to help understand the usefulness of the estimates made. Therefore, when one speaks of an *optimum concentration*, one is really dealing with an optimum range, rather than a single point. Very few researchers follow that practice.

The various elements are taken up in ionic form, some of which are complementary to each other, others are competitive. When essential elements are applied to soils, the uptake by the plant will be interactive and reactive with the soil, soil microorganisms, weeds, and other plants that may be competing for these elements (diseases, nematodes, and insects that infect or feed on roots can also decrease the uptake and utilization of essential elements present in soils); this is why a properly applied foliar application of some elements, particularly the micronutrients, is often much more efficient than soil applications, but may have an adverse impact on the interpretation of plant analysis results.

Munson and Nelson (1990) also discussed the use of plant analysis nutrient ratios or the Diagnosis and Recommendation Integrated System (DRIS), a comprehensive system initially proposed by Beaufils (1961, 1973). As initially proposed, the system was to include soil nutrient levels, plant analysis, management (e.g., irrigation, plant population, row widths, etc.) and cultural practices (such as tillage) as well as climatic or geographic information. The plant analysis ratios would cancel out the dry-matter factor, and therefore were thought to give a wider coverage, making age of the plant or stage of crop development of less importance. The use of DRIS has been studied and reviewed by Sumner (1974), Walworth and Sumner (1987), and Beverly (1991). Black (1993) also has a comprehensive review of the use of plant analysis nutrient ratios, including DRIS, and the calculation of nutrient indices. Most of the studies to date have involved yield levels and plant analysis values for different elements and have not included measured items as comprehensively as initially proposed by Beaufils.

Dow and Roberts (1982) developed the concept of seasonal critical nutrient ranges and Roberts and Dow (1982) discussed the advantage of using the 95 and 100% yield regression equations relating percent petiole P in potato to establish seasonal deficient, critical nutrient range, and adequate nutrient concentration zones for crops so that plant analysis sampling can take place over the season in order that a proper diagnosis or interpretation can be made (Figure 1.4). This approach needs to be expanded and applied simultaneously for several elements for meaningful interpretation.

INTERPRETATION

Interpretation of the results of plant analysis is very important. Among those that have written on interpretation of results are Goodall and Gregory (1947), Ulrich (1948), Childers (1966), Jones (1967), Chapman (1966; 1967), Reuter and Robinson (1986), Jones et al. (1991), and Mills and Jones (1996) for crops in general; Ulrich and Hills (1990) for sugar beets; Bowen (1990) for sugarcane; Sabbe and Zelinski (1990) for cotton; Westfall et al. (1990) for small grains; Jones et al. (1990) for corn and sorghum; Frazier et al. (1967) and Small and Ohlrogge (1973) for soybeans; Geraldson and Tyler (1990) for vegetable crops; Kenworthy (1967, 1973) and Righetti et al. (1990) for fruits and orchards; Kelling and Matocha (1990) for forage crops; and Weetman and Wells (1990) for forests. Bergmann (1992) has presented nutrient ranges for 10 elements for a host of crops. These references can be very helpful for those looking for information on specific crops and can provide insights as to how these authors interpret plant analyses for those crops.

The general steps one needs to follow when establishing critical levels or indices for elements determined in a plant analysis include:

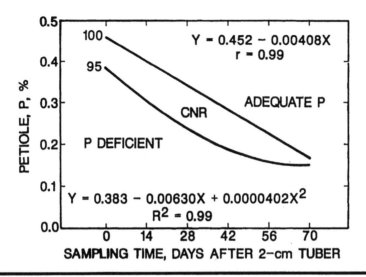

FIGURE 1.4 Use of the 95 and 100% of maximum potato yield regression equations, from percent petiole P and sampling time in days after the formation of 2-cm tubers to establish seasonal deficient, critical nutrient range (CNR), and adequate zones (Roberts and Dow, 1982; Munson and Nelson, 1990). Used with permission of the Soil Science Society of America, Madison, WI.

1. Conduct controlled experiments with adequate treatments over a period of time and conditions that will yield reliable nutrient levels to establish or define the zones or levels.
2. Obtain reliable plant samples (Jones and Case, 1990). To do that one must determine:
   a. The portion of plant sampled.
   b. The number of samples needed to represent the management unit.
   c. The time of sampling or best physiological stage relative to diagnostic norms.
3. Handle samples properly (Jones and Case, 1990).
   a. Identify samples with respect to location and number.
   b. Decontaminate samples (Jones and Case, 1990).
   c. Store and handle samples properly for transit.
   d. Dry and properly handle samples before grinding.
   e. Grind samples using non-contaminating mills and sieves.
   f. Store ground samples securely before and after analysis so that follow-up analyses can be made if needed.

4. Use accepted methods of laboratory preparation and analysis on the samples.
5. Record resulting data for transmittal or transmission to user (interpretative information may accompany the data, depending on the experience and skill of the user).

PRACTICAL USES OF PLANT ANALYSIS

Plant analysis results are used in a number of ways:

1. Determining that an element is essential for plant growth, development, and maturation or some process closely associated with nutrition and growth, such as symbiotic N fixation
2. Verifying the element associated with a phenotypic or apparent nutrient element deficiency or toxicity symptom
3. Establishing optimum concentrations or critical values for elements associated with optimum or maximum economic yields
4. Determining availability of soil nutrients and/or availability and recovery of an applied element or elements in fertilizer in crop response experiments
5. Evaluating and comparing different areas or sites within a production field, orchard, or forest
6. Monitoring fields, vineyards, orchards, or forests to determine if deficiencies or nutrient imbalances exist so that corrective action may be taken if deemed necessary
7. Determining the total elemental uptake by a crop, particularly at high levels of production (by this process, an estimate of the nutrient element requirement per unit of production can be determined as well as nutrient element removal per unit of yield)
8. Determining the internal nutrient efficiencies (output per unit of nutrient uptake) of varieties or cultivars based on the recovery of soil and/or fertilizer nutrients by different methods of application, including foliar
9. Conducting field surveys of crops grown in production or soil areas to determine if an element or a combination of elements is or are deficient, sufficient, in excess, or toxic

Munson (1992) has given the results of the total nutrient uptake by high-yielding crops found by various researchers. The total nutrient uptake by corn (maize) yielding 19.3 and 21.2 t/ha of grain, soybean at 5.38 and 6.79 t/ha, potato at 66.8 t/ha, alfalfa at 24.1 t/ha, and various species of wheat from 3.79 to over 11 t/ha are given. For the higher-yield levels, total uptake of N, P, and K were 434 kg N, 94.8 kg P, and 424 kg K for corn; for soybean, 614 kg N,

64.6 kg P, and 402 kg K; for potato, 140 kg N, 28.8 kg P, and 269 kg K; for alfalfa, 731 kg N, 66 kg P, and 532 kg K; and for wheat, 261 kg N, 42.8 kg P, and 247 kg K. Secondary and micro-nutrient data are presented for some of the crops. Some variations in nutrient element uptake were found, depending on whether or not the crop was sampled at the peak uptake period for each of the elements determined.

From seasonal sampling, one can determine the peak rates of demand for a specific nutrient element at different stages of growth for different crops (Munson and Nelson, 1990). Recent studies by Sadler and Karlen (1995) on five different soybean experiments indicate that soybean varieties as well as experimental and weather factors appear to influence the rates of elemental uptake differently than those found for corn and wheat, suggesting the need for further study.

From corn leaves sampled at pollination under very high levels of production, average results give an indication of what the optimum levels of nutrient concentration might be, levels reported by Flannery (1986) for a 5-year average yield of 19.2 t/ha and Lockman (1986) for a 3-year average yield of 19.5 t/ha. A comparison of the average leaf element concentrations from widely divergent production areas, New Jersey and central Illinois, is shown in Table 1.5, levels that are obviously within the adequate zone or sufficiency range for all of these elements.

Some researchers have found that one can analyze grain or seeds to determine if the availability of an element is optimum for producing the maximum yield. Pierre et al. (1977a,b) studied the N concentration in corn grain versus yield, expressed as a percentage of the maximum, as a measure of N sufficiency. Using both the regression and graphic analysis techniques, they concluded that as corn yields approached the maximum yield, the post-harvest grain contained 1.54 and 1.52% N as determined by the respective techniques.

TABLE 1.5 Five- and three-year average corn yields and elemental ear leaf concentrations sampled at early tassel from experiments in New Jersey and Illinois

Average grain yield (t/ha)	*Average elemental concentration: ear leaf*									
	%						*ppm*			
	N	*P*	*K*	*Ca*	*Mg*	*S*	*B*	*Cu*	*Mn*	*Zn*
19.2[a]	3.14	0.36	2.51	0.56	0.18	0.22	13	10	39	30
19.5[b]	3.37	0.30	2.70	0.51	0.14	0.27	7.7	5.7	48	27

[a] New Jersey.
[b] Illinois.

TABLE 1.6   Concentration of nutrients in soybean seed at two yield levels and row widths (Bundy and Oplinger, 1984)

Soybean yield (t/ha	Row width (cm)	\multicolumn{11}{c}{Nutrient concentration in the seed}

Soybean yield (t/ha	Row width (cm)	N	P	K	Ca	Mg	S	B	Cu	Mn	Zn
		\multicolumn{6}{c}{%}	\multicolumn{3}{c}{ppm}		
5.046	20.3	5.85	0.60	1.79	0.19	0.24	0.31	41	25	26	41
4.435	76.2	5.83	0.60	1.80	0.20	0.23	0.30	43	25	27	43

Lockman (1986) found that corn grain contained 1.6% N from a yield level of over 20.4 t/ha, which was higher than that obtained by Pierre et al. (1977a,b). Goos et al. (1982) studied grain protein of winter wheat as a post-harvest measure of the adequacy of N sufficiency for maximum grain yield. They concluded that the critical grain protein level was 11.5%, and that the transition zone between deficiency and sufficiency was from 11.1 to 12.0%.

The elemental content in some seeds, such as soybean, appears to have a certain stability. For example, researchers in Wisconsin have found that by changing a management practice, such as narrowing the row spacing, yield could be increased over 1.4 t/ha with little change in the elemental content of the beans produced (Table 1.6). From such data, one can readily obtain the nutrient element removal based on the dry matter in the harvested product.

Nutrient element removal by high-yielding crops is much greater than many believe. For example, Lockman (1986) analyzed grain each year from corn produced on an Illinois farm that had a 3-year average yield of over 19.5 t/ha (312 bu/a). The average grain elemental content was N = 1.57%; P = 0.32%; K = 0.31%; Ca = 0.30%; Mg = 0.12%; S = 0.13%; B = 1.6 ppm; Cu = 3.1 ppm; Fe = 30 ppm; Mn = 8 ppm; and Zn = 41 ppm. For the major fertilizer elements, N, P, and K, the average annual removals in the grain would have been over 250, 53, and 51 kg/ha, respectively, a ratio of about 5:1:1. In terms of P_2O_5 and K_2O as fertilizer, the amounts would be over 119 kg P_2O_5/ha and 61 kg K_2O/ha, an N:P:K ratio of about 4.2:1.9:1. This, of course, does not indicate the rate of application for these fertilizer elements, but merely demonstrates what removal levels occur at such grain yields. If the corn plants had been harvested as silage, K removals would have been much higher. Flannery (1986) found that for corn production levels at 21.2 t grain/ha, total K removal amounted to 424 kg K/ha when the entire crop was harvested. Such high removals can rather quickly draw down soil levels unless careful soil and plant diagnostic accounting methods are followed. Most forage crops, when harvested for either hay or haylage, remove very high quantities of nutrients.

SUMMARY

Plant analysis can play a major role when diagnosing mineral nutrition problems, whether for research purposes or for solving practical field problems for farmers and growers. However, the application of plant analysis techniques requires skill and experience on the user's part. Most of the essential elements required by crops and the climatic, genetic, cultural, and management factors that influence crop growth, development, and maturation must be factored into the interpretation of a plant analysis result and a resulting recommendation. All of these factors are interactive. When only one essential element is lacking or inadequate rainfall or irrigation is not provided at a crucial time in the life cycle of the crop, the crop yield may drop to zero.

Another aspect might be the manner in which the concentration levels for sufficiency are determined relative to the maximum yield of the crop. For example, if only one element is considered, the sufficiency level of that element being at a concentration that produced 90% of maximum yield, the true production potential of that crop may not be achieved because of the other 12 elements (excluding water) that are interacting with that element. Based on the percent sufficiency concept, it has been proposed that the elements have a multiplicative effect on yield. For example, if only two elements were at concentrations that produce 90% of the maximum yield, and the other 11 are at concentrations that produce 100% of the maximum yield, only 81% of the maximum would be achieved due to the multiplicative effect. If levels of 90% sufficiency were also selected for all 13 elements, theoretically only 25.4% of the maximum yield would be obtained. In the author's view, this multiplicative, interactive effect of the elements on yield has yet to be thoroughly evaluated, although the concept is worthy of further study.

Over the last several decades, crop yields in general have continued to increase, with plant nutrition through proper fertilization playing a major role in these increases that have occurred throughout the world (Halliday and Trenkel, 1992). Plant analysis, with improving diagnostic interpretation, has also played a key role in those increases.

While plant analysis is not the final answer with respect to regulating the mineral nutrition of crop plants, it has been and will continue to be one of the most useful tools leading to improved crop production and quality. When used in conjunction with other diagnostic techniques by farmers, consultants, agricultural dealers, or extension personnel, a plant analysis result can be extremely helpful in both diagnostic and monitoring roles. The real test of the value of a plant analysis result, however, comes from the solving of a practical problem faced by a grower or when it contributes to the analysis and interpretation of research results.

The information presented in this chapter, and the others that follow, will strengthen the reader's understanding of the plant analysis technique, and aid in using the technique in order to support the goal of improving agricultural production and profitability.

REFERENCES

Adriano, D.C. 1986. *Trace Elements in the Terrestrial Environment.* Springer-Verlag, New York.

Ahmed, S. and H.J. Evans. 1960. Cobalt: A micronutrient element for the growth of soybean plants under symbiotic conditions. *Soil Sci.* 90:205–210.

Arnon, D.I. and P.R. Stout. 1939. The essentiality of certain elements in minute quantity for plants with special reference to copper. *Plant Physiol.* 14:371–375.

Asher, C.J. 1991. Beneficial elements, functional nutrients, and possible new essential elements, pp. 703–723. In: J.J. Mortvedt et al. (Eds.), *Micronutrients in Agriculture.* Second edition. SSSA Book Series 4. Soil Science Society of America, Madison, WI.

Bear, F.E. 1948. Historic introduction, pp. ix–xxiii. In: H.B. Kitchen (Ed.), *Diagnostic Techniques for Soils and Crops.* The American Potash Institute, Washington, D.C.

Beaufils, E.R. 1961. Les desequilibres chimiques chez *l'Hevea brasiliensis.* La methode dite du diagnostic physiologique. Thesis, Paris University (Sorbonne), France.

Beaufils, E.R. 1973. *Diagnostic and Recommendation Integrated System (DRIS).* Soil Science Bulletin No. 1, Department of Soil Science and Agrometeorology, University of Natal, Pietermaritzburg, South Africa.

Bennett, W.F. (Ed.). 1993. Nutrient Deficiencies and Toxicities in Crop Plants. APS Press, The American Phytopathological Society, St. Paul, MN.

Bergmann, W. 1992. *Nutritional Disorders of Plants: Development, Visual and Analytical Diagnosis.* Gustav Pischer Verlag, Jena, Germany.

Beverly, R.B. 1991. A Practical Guide to the Diagnosis and Recommendation Integrated System (DRIS). Micro-Macro Publishing, Athens, GA.

Black, C.A. 1993. *Soil Fertility Control and Evaluation.* Lewis Publishers, Boca Raton, FL.

Bowen, J.E. 1990. Plant tissue analysis of sugarcane, pp. 449–467. In: R.L. Westerman (Ed.), *Soil Testing and Plant Analysis.* SSSA Book Series 3. Soil Science Society of America, Madison, WI.

Brown, P.H., R.M. Welsh, and E.E. Cary. 1987. Nickel: A micronutrient essential for higher plants. *Plant Physiol.* 85:801–803.

Bundy, L.G. and E.S. Oplinger. 1984. Narrow row spacings increase soybean yields and nutrient removal. *Better Crops With Plant Food* 68:16–17.

Chapman, H.D. (Ed.). 1966. *Diagnostic Criteria for Plants and Soils.* University of California, Division of Agriculture, Berkeley.

Chapman, H.D. 1967. Plant analysis values suggestive of nutrient status of selected crops, pp. 77–92. In: G.W. Hardy (Ed.), *Soil Testing and Plant Analysis. Plant Analysis.* Part II. Special Publication No. 2. Soil Science Society of America, Madison, WI.

Childers, N.F. 1966. *Nutrition of Fruit Crops, Temperate, Sub-tropical, Tropical.* Horticultural Publications, Rutgers—The State University, New Brunswick, NJ.

Dalton, D.A., H.J. Evans, and P.H. Hanus. 1985. Stimulation by nickel of soil microbial urease activity, and hydrogenase activities in soybeans grown in low-nickel soil. *Plant Soil* 88:244–258.

De Saussure, N.T. 1804. *Recherches chimiques sur la vegetation.* Nyon. Paris, France.

Dow, A.I. and S. Roberts. 1982. Proposal: Critical nutrient ranges for crop diagnosis. *Agron. J.* 74:401–403.

Epstein, E. 1965. Mineral nutrition, pp. 438–466. In: J. Bonner and J.E. Verner (Eds.), *Plant Biochemistry.* Academic Press, New York, NY.

Flannery, R.L. 1986. Plant food uptake in maximum yield corn study. *Better Crops With Plant Food* 70:4–5.

Frazier, R.D., H.G. Small, and A.J. Ohlrogge. 1967. Nutrient concentrations in plant parts sampled from soybean fields, pp. 33–47. In: G.W. Hardy (Ed.), *Soil Testing and Plant Analysis.* Plant Analysis. Part II. Special Publication No. 2. Soil Science Society of America, Madison, WI.

Geraldson, C.M. and K.B. Tyler. 1990. Plant analysis as an aid in fertilizing vegetable crops, pp. 549–562. In: R.L. Westerman (Ed.), *Soil Testing and Plant Analysis.* SSSA Book Series 3. Soil Science Society of America, Madison, WI.

Glass, A.D.M. 1989. *Plant Nutrition: An Introduction to Current Concepts.* Jones and Bartlett Publishers, Boston, MA.

Goodall, D.W. and P.G. Gregory. 1947. *Chemical Composition of Plants as an Index of their Nutritional Status.* Imperial Bureau Horticultural Plantation Crops. Technical Communication No. 17. Ministry of Agriculture, London, England.

Goos, R.J., D.G. Westfall, A.E. Ludwick, and J.E. Goris. 1982. Grain protein content as an indicator of N sufficiency for winter wheat. *Agron. J.* 74:130–133.

Grundon, N.J. 1987. *Hungry Crops: A Guide to Nutrient Deficiencies in Field Crops.* Queensland Department of Primary Industries, Brisbane, Australia.

Halliday, D.J. and M.E. Trenkel (Eds.). 1992. *IFA World Fertilizer Use Manual.* International Fertilizer Industry Association, Paris, France.

Hardy, G.W. (Ed.). 1967. *Soil Testing and Plant Analysis. Plant Analysis.* Part II. Special Publication No. 2. Soil Science Society of America, Madison, WI.

Jones, Jr., J.B. 1967. Interpretation of plant analysis for several agronomic crops, pp. 49–58. In: G.W. Hardy (Ed.), *Soil Testing and Plant Analysis.* Plant Analysis. Part II. Special Publication No. 2. Soil Science Society of America, Madison, WI.

Jones, Jr., J.B. 1994a. *Plant Nutrition Manual.* Micro-Macro Publishing, Athens, GA.

Jones, Jr., J.B. 1994b. Plant Analysis (VHS video). St. Lucie Press, Delray Beach, FL.

Jones, Jr., J.B. 1994c. Tissue Testing (VHS video). St. Lucie Press, Delray Beach, FL.

Jones, Jr., J.B. and V.W. Case. 1990. Sampling, handling, and analyzing plant tissue samples, pp. 389–427. In: R.L. Westerman (Ed.), *Soil Testing and Plant Analysis.* SSSA Book Series 3. Soil Science Society of America, Madison, WI.

Jones, Jr., J.B., H.V. Eck, and R. Voss. 1990. Plant analysis as an aid in fertilizing corn and grain sorghum, pp. 521–547. In: R.L. Westerman (Ed.), *Soil Testing and Plant Analysis*. SSSA Book Series 3. Soil Science Society of America, Madison, WI.

Jones, Jr., J.B., B. Wolf, and H.A. Mill s. 1991. *Plant Analysis Handbook: A Practical Sampling, Preparation, Analysis, and Interpretation Guide*. Micro-Macro Publishing, Athens, GA.

Kabata-Pendias, A. and H. Pendias. 1995. *Trace Elements in Soils and Plants*. Second edition. CRC Press, Boca Raton, FL.

Kelling, K.A. and J.E. Matocha. 1990. Plant analysis as an aid in fertilizing forage crops, pp. 603–643. In: R.L. Westerman (Ed.), *Soil Testing and Plant Analysis*. SSSA Book Series 3. Soil Science Society of America, Madison, WI.

Kenworthy, A.L. 1967. Plant analysis and interpretation of analysis for horticulture crops, pp. 59–75. In: G.W. Hardy (Ed.), *Soil Testing and Plant Analysis*. Plant Analysis, Part II. Special Publication No. 2. Soil Science Society of America, Madison, WI.

Kenworthy, A.L. 1973. Leaf analysis as an aid in fertilizing orchards, pp. 381–392. In: L.M. Walsh and J.D. Beaton (Eds.), *Soil Testing and Plant Analysis*. Revised edition. Soil Science Society of America, Madison, WI.

Krantz, B.A., W.L. Nelson, and L.P. Burkhart. 1948. Plant-tissue tests as a tool in agronomic research, pp. 137–155. In: H.B. Kitchen (Ed.), *Diagnostic Techniques for Soils and Crops*. The American Potash Institute, Washington, D.C.

Lockman, R.B. 1986. Personal communication. Agrico Chemical Company Testing Laboratory. Washington Court House, OH.

Macy, P. 1936. The qualitative mineral nutrient requirements of plants. *Plant Physiol.* 11:749– 764.

Market, B. 1992. Presence and significance of naturally occurring chemical elements of the periodic system in the plant organism and consequences for future investigations on inorganic environmental chemistry in ecosystems. *Vegetation* 1203:1–30.

Marschner, H. 1986. *Mineral Nutrition of Higher Plants*. Academic Press, New York.

Martin-Prével, J. Gagnard, and P. Gauier. 1987. *Plant Analysis: As a Guide to the Nutrient Requirements of Temperate and Tropical Crops*. Lavoisier Publishing, New York.

McMurtrey, J.E., Jr. 1948. Visual symptoms of malnutrition in plants, pp. 231–209. In: H.B. Kitchen (Ed.), *Diagnostic Techniques for Soils and Crops*. The American Potash Institute, Washington, D.C.

Mengel, K. and E.A. Kirkby. 1982. Elements with more toxic effects, pp. 559–571. In: *Principles of Plant Nutrition*. Third edition. International Potash Institute, Bern, Switzerland.

Mills, H.A. and J.B. Jones, Jr. 1996. *Plant Analysis Handbook II*. Micro-Macro Publishing, Athens, GA.

Morgan, N.D. and G. A. Wickstrom. 1956. Give your plants a blood test: Guide to quick tissue tests. *Better Crops With Plant Food*. Reprint L-5-56.

Munson, R.D. 1992. Potassium in ecosystems, with emphasis on the U.S., pp. 279–308. In: *Potassium in Ecosystems, Biogeochemical Fluxes of Cations in Agro- and Forest*

Systems. Proceedings of the 23rd Colloquium of the International Potash Institute, held at Prague, Czechoslovakia. International Potash Institute, Basel, Switzerland.

Munson, R.D. and W.L. Nelson. 1973. Principles and practices in plant analysis, pp. 223–248. In: L.M. Walsh and J.D. Beaton (Eds.), *Soil Testing and Plant Analysis.* Revised edition. Soil Science Society of America, Madison, WI.

Munson, R.D. and W.L. Nelson. 1990. Principles and practices in plant analysis, pp. 359–387. In: R.L. Westerman (Ed.), *Soil Testing and Plant Analysis.* SSSA Book Series 3. Soil Science Society of America, Madison, WI.

Ohlrogge, A.J. 1962. *The Purdue Soil and Plant Tissue Tests.* Purdue University Station Bulletin 635 (Rev.). Purdue University, West Lafayette, IN.

Pais, I. 1992. Criteria of essentiality, beneficiality, and toxicity of chemical elements. *Acta Aliment.* 21(2):145–152.

Pais, I. and J.B. Jones, Jr. 1996. *Handbook on Trace Elements in the Environment.* St. Lucie Press, Delray Beach, FL.

Pierre, W.H., V.D. Jolley, J.R. Webb, and W.D. Schrader. 1977a. Relation between corn yield, expressed as percentage of the maximum, and the N percentage in the grain. I. Various N-rate experiments. *Agron. J.* 69:215–220.

Pierre, W.H., L. Dumenil, and J. Henao. 1977b. Relationship between corn yield, expressed as percentage of the maximum, and the N percentage of the grain. II. Diagnostic use. *Agron. J.* 69:221–226.

Reuter, D.J. and J.B. Robinson (Eds.). 1986. *Plant Analysis: An Interpretation Manual.* Inkata Press Pty. Ltd., Melbourne, Australia.

Righetti, T.L., K.L. Wilder, and G.A. Cummings. 1990. Plant analysis as an aid in fertilizing orchards, pp. 563–601. In: R.L. Westerman (Ed.), *Soil Testing and Plant Analysis.* SSSA Book Series 3. Soil Science Society of America, Madison, WI.

Risser, J.A. and D.E. Baker. 1990. Soil testing for toxic metals, pp. 275–298. In: R.L. Westerman (Ed.), *Soil Testing and Plant Analysis.* SSSA Book Series 3. Soil Science Society of America, Madison, WI.

Roberts, S. and A.I. Dow. 1982. Critical nutrient ranges for petiole phosphorus levels of sprinkler-irrigated Russet Burbank potatoes. *Agron. J.* 74:583–585.

Sabbe, W.E. and L.J. Zelinski. 1990. Plant analysis as an aid in fertilizing cotton, pp. 469–493. In: R.L. Westerman (Ed.), *Soil Testing and Plant Analysis.* SSSA Book Series 3. Soil Science Society of America, Madison, WI.

Sadler, E.J. and D.L. Karlen. 1995. Aerial dry matter and nutrient accumulation comparisons among five soybean experiments. *Commun. Soil Sci. Plant Anal.* 26:3145–3163.

Small, Jr., H.G. and A.J. Ohlrogge. 1973. Plant analysis as an aid in fertilizing soybeans and peanuts, pp. 315–327. In: L.M. Walsh and J.D. Beaton (Eds.), *Soil Testing and Plant Analysis.* Revised edition. Soil Science Society of America, Madison, WI.

Smith, P.F. 1962. Mineral analysis of plant tissues. *Annu. Rev. Plant Physiol.* 13:81–108.

Sprague, H B. (Ed.). 1964. *Hunger Signs in Crops.* Third edition. David McKay Company, New York, NY.

Steenbjerg, P. 1951. Yield curves and chemical plant analyses. *Plant Soil* 3:97–109.

Syltie, P.W., S.W. Melsted, and W.M. Walker. 1972. Rapid tissue tests as indicators of yield, plant composition, and fertility for corn and soybeans. *Commun. Soil Sci. Plant Anal.* 3:37–49.

Sumner, M.E. 1974. An evaluation of Beaufil's physiological diagnosis technique for determining the nutrient requirement of crops, pp. 437–446. In: J. Wehrmann (Ed.), *Proceedings 7th International Colloquium on Plant and Fertilizer Problems.* Volume 2. German Society of Plant Nutrition, Hanover, Germany.

Takahashi, E., J.F. Ma, and Y. Miyake. 1990. The possibility of silicon as an essential element for higher plants, pp. 99–102. In: *Comments on Agriculture and Food Chemistry.* Gordon and Beach Scientific Publishers, London, England.

Ulrich, A. 1948. Plant analysis methods and interpretation of results, pp. 157–198. In: H.B. Kitchen (Ed.), *Diagnostic Techniques for Soils and Crops.* The American Potash Institute, Washington, D.C.

Ulrich, A. and P.J. Hills. 1967. Principles and practices of plant analysis, pp. 11–24. In: G.M. Hardy (Ed.), *Soil Testing and Plant Analysis.* Plant Analysis, Part II. Special Publication No. 2. Soil Science Society of America, Madison, WI.

Ulrich, A. and P.J. Hills. 1990. Plant analysis as an aid in fertilizing sugar beets, pp. 429–447. In: R.L. Westerman (Ed.), *Soil Testing and Plant Analysis.* SSSA Book Series 3. Soil Science Society of America, Madison, WI.

von Liebeg, J. 1840. Organic chemistry in its application to agriculture and physiology (Edited from notes to the author by Lynn Play Fair). Taylor and Walton, London, England.

Walsh, L.M. and J.D. Beaton (Eds.). 1973. *Soil Testing and Plant Analysis.* Revised edition. Soil Science Society of America, Madison, WI.

Walworth, J.L. and M.E. Sumner. 1987. The Diagnosis and Recommendation Integrated System (DRIS), pp. 149–188. In: B.A. Stewart (Ed.), *Advances in Soil Science.* Volume 6. Springer-Verlag, New York.

Weetman, G.F. and C.G. Wells. 1990. Plant analysis as an aid in fertilizing forests, pp. 659–690. In: R.L. Westerman (Ed.), *Soil Testing and Plant Analysis.* SSSA Book Series 3. Soil Science Society of America, Madison, WI.

Weinhold, A. 1862. Analyze von unkiauterm des bodens der versuchsstation chemnitz. *Landeo Vers. Sta.* 4:188–193.

Westerman, R.L. (Ed.). 1990. *Soil Testing and Plant Analysis.* SSSA Book Series 3. Soil Science Society of America, Madison, WI.

Westfall, D.G., D.A. Whitney, and D.M. Brandon. 1990. Plant analysis as an aid in fertilizing small grain, pp. 495–519. In: R.L. Westerman (Ed.), *Soil Testing and Plant Analysis.* SSSA Book Series 3. Soil Science Society of America, Madison, WI.

Wickstrom, G.A. 1963. Ask the plant about N-P-K needs. *Solutions,* May-June:36–37.

Wickstrom, G.A. 1967. Use of tissue testing in field diagnosis, pp. 109–112. In: G.M. Hardy (Ed.), *Soil Testing and Plant Analysis.* Plant Analysis, Part II. Special Publication No. 2. Soil Science Society of America, Madison, WI.

FIELD SAMPLING PROCEDURES FOR CONDUCTING A PLANT ANALYSIS

2

J. Benton Jones, Jr.

INTRODUCTION

The validity and usefulness of the determined elemental content of a collected plant tissue sample hinge on an intelligent and realistic approach to the problem of how to obtain a reliable sample. If the sample taken is not representative of the general population, all the careful and costly work put into the subsequent analysis will be wasted because the results will be invalid. To obtain a representative sample from a particular plant species is a complex problem, and expert knowledge is required before it can be attempted.

ELEMENTAL HETEROGENEITY

The elemental content of a plant is not a fixed entity, but varies from month to month, day to day, and even from hour to hour, as well as differing between the various parts of the plant itself (Goodall and Gregory, 1947; Jones, 1970). A plant part at a specific location on the plant obtained at a definite stage of growth (on the basis of physiological age) constitutes the sampling parameters.

1-57444-124-8/98/$0.00+$.50
© 1998 by CRC Press LLC

In general, tissues that are either physiologically young and undergoing rapid change in elemental content or those past full maturity should not be sampled.

The plant part selected and the time of sampling must correspond to the best relationship that exists between its elemental content and yield or the physical appearance of the plant (Bates, 1971). Frequently, no single time for sampling of a particular plant part is ideal for evaluating every element; therefore, several plant parts at different growth stages may need to be taken. Comparisons of analyses between leaves and petioles, stems and leaves, and upper and lower plant parts may assist in evaluating a plant analysis (Bates, 1971). Jones (1967) noted that the determination of the homogeneity, or the lack of it, may be a useful technique when diagnosing certain suspected elemental deficiencies in some crops. For example, corn (*Zea mays* L.) plants deficient in potassium (K) contain less K in their lower leaves than in their upper leaves. When plants contain sufficient K, the reverse is true. It was also noted that differences in concentrations of boron (B) and zinc (Zn) among the upper and lower leaves of corn decrease as the plant approaches B or Zn deficiency (Jones, 1967). However, the practical application of this technique of comparison of analysis results between plant parts has yet to develop into a practical system of plant analysis interpretation.

STATISTICAL CONSIDERATIONS

Once it has been determined which plant part is to be sampled to represent the plant's elemental status, the number of plants to sample for adequate representation must be decided. What constitutes an adequate number has been determined to some degree from the results of previous research.

Plants growing adjacent to each other can differ considerably in their elemental content. Lilleland and Brown (1943), when studying the phosphorus (P) nutrition of peach (*Prunus persica* L.) trees, found that the composition of morphologically homologous leaves taken from adjacent trees receiving the same fertilizer treatment differed considerably. This was also the experience that Thomas (1945) found with apple (*Malus sylvestris* Mill.) trees, and Steyn (1959) with citrus trees and pineapple [*Ananas comosus* (L.) Merr.] plants.

Elemental content variations within single plant parts as well as from plant to plant must be considered. Steyn (1959) has shown that there are relatively small variations in the nitrogen (N), phosphorus (P), K, calcium (Ca), magnesium (Mg), iron (Fe), manganese (Mn), Zn, and copper (Cu) concentrations among the selected sampling material on a single citrus tree or a single pineapple plant. It is reasonable to assume that this could be the case for other plants as well. Therefore, provided the sampled material is carefully selected, a relatively small sample could adequately represent the elemental content of a single plant.

When considering the variation in element content from plant to plant, the situation can be entirely different. If large variations exist, intensive sampling is required to obtain sufficient plant tissue for representing the element content of the plant sampled. Steyn (1961) conducted a statistical sampling study on citrus trees and pineapple plants in which adjacent plants in blocks were intensively sampled under rigorously controlled conditions. Some of the analysis results from the citrus trees show the minimum number of trees (in blocks of 16 trees) to sample to obtain 85% level of significance in differences (D) exceeding D% of the mean values:

| | *Element* | | | | | | | | |
| | *g/kg* | | | | | *mg/kg* | | | |
Item	*N*	*P*	*K*	*Ca*	*Mg*	*Fe*	*Mn*	*Zn*	*Cu*
Mean	26.8	1.17	4.8	36.3	4.8	80.0	23.4	11.0	5.9
CV (%)	5.3	3.6	15.8	7.2	18.3	8.0	16.9	20.0	18.3
					Number of trees				
D = 10%	3	2	23	5	31	6	26	37	31
D = 20%	1	1	6	2	8	2	7	10	8

It was found that, when the citrus trees were in poor condition, much more intensive sampling than that mentioned previously was necessary for adequate representation. In the Steyn (1961) study, variations in N and P content were usually considerably less than that observed for the other elements. Potassium and Mg showed the greatest degree of variation, followed by some of the micronutrients, such as Cu and Zn. When the concentration of the element was at a deficient level, the variation was exceptionally large. It is evident that for Zn virtually all the trees in the block of 16 would have to be sampled for adequate representation of the plant Zn status. Similar results were obtained in the pineapple sampling study. Therefore, if all the essential elements are to be determined in a single sample, a requirement to adequately interpret a plant analysis is to follow an intensive sampling procedure. As was observed for citrus, many leaf tissue samples would be required if pineapple plants under nutrient element stress were being evaluated for their elemental content.

Colonna (1970), working in a homogeneous coffee (*Coffea arabica* L.) plantation, recommended sampling 2 leaves per tree from 40 randomly selected trees per hectare in 1 hectare. In fertilizer trials, he found it necessary to sample 5 to 6 replicates with 20 to 25 coffee plants per treatment to obtain useful plant analysis data. Similar sampling studies for the more commonly grown annual field crops have yet to be done to establish the sampling intensity required to ensure reasonable analytical reliability.

PLANT SAMPLING TECHNIQUES

Kenworthy (1969), Chapman (1966), Jones et al. (1971, 1973), Reuter and Robinson (1986, 1997), Jones, Wolf, and Mills (1993), and Mills and Jones (1996) have described plant tissue sampling techniques that have been generally accepted. The following is a partial list of recommended sampling procedures taken from these sources:

Suggested sampling procedures for field and vegetable crops, fruits and nuts, and ornamental plants

Crop	Stage of growth	Plant part to sample	Number of plants to sample
Field crops			
Corn	Seedling stage (<12 in.)	All the aboveground portion	20–30
	Prior to tasseling	The entire leaf fully developed below the whorl	15–25
	From tasseling and shooting to silking	The entire leaf at the ear node (or immediately above or below it)	15–25
Soybean or	Seeding stage (<12 in.)	All the aboveground portion	20–30
other bean	Prior to or during flowering[a]	Two or three fully developed leaves at the top of the plant.	20–30
Small grains (including rice)	Seedling stage (<12 in.)	All the aboveground portion	50–100
	Prior to heading	The fourth uppermost leaves	50–100
Hay, pasture or forage grasses	Prior to seed head emergence or ae the optimum stage for best quality forage	The fourth uppermost leaf blades	40–50

Suggested sampling procedures for field and vegetable crops, fruits and nuts, and ornamental plants (continued)

Crop	Stage of growth	Plant part to sample	Number of plants to sample
Alfalfa	Prior to or at 1/10 bloom stage	Mature leaf blades taken about one-third of the way down the plant	40–50
Clover and other legumes	Prior to bloom	Mature leaf blades taken about one- third of the way down the plant	40–50
Sugar beets	Mid-season	Fully expanded and mature leaves midway between the younger center leaves and the oldest leaf whorl on the outside	40–50
Tobacco	Before bloom	Uppermost fully developed leaf	8–12
Sorghum-milo	Prior to or at heading	Second leaf from top of plant	15–25
Peanuts	Prior to or at bloom stage	Mature leaves from both the main stem and either cotyledon lateral branch	40–50
Cotton	Prior to or at first bloom or when squares appear	Youngest fully mature leaves on main stem squares appear	30–40
Vegetable crops			
Potato	Prior to or during early bloom	Third to sixth leaf from growing tip	20–30
Head crops (cabbage, etc.)	Prior to heading	First mature leaves from center of the whorl	10–20
Tomato (field)	Prior to or during early fruit set	Third or fourth leaf fromp growing tip	20–25
Tomato (greenhouse)	Prior to or during fruit set	Young plants: leaves adjacent to second and third clusters	20–25
		Older plants: leave from fourth to sixth clusters	20–25
Bean	Seedling stage (<12 in.)	All the aboveground portion	20–30
	Prior to or during initial flowering	Two or three fully developed leaves at the top of the plant	
Root crops (carrots, onions, beets, etc.)	Prior to root or buld and enlargement	Center mature leaves	20–30

Suggested sampling procedures for field and vegetable crops, fruits and nuts, and ornamental plants (continued)

Crop	Stage of growth	Plant part to sample	Number of plants to sample
Vegetable crops			
Celery	Mid-growth (12 to 15 in. tall)	Petiole of youngest mature leaf	15–30
Leaf crops (lettuce, spinach, etc.)	Mid-growth flowering	Youngest mature leaf from the top of the plant	35–60
Peas	Prior to or during initial flowering	Leaves from the third node down	30–60
Sweet corn	Prior to tasseling	The entire fully mature leaf below the whorl	
	At tasseling	The entire leaf at the ear node	20–30
Melons (watermelon, cucumber, muskmelon)	Early stages of growth prior to fruit set	Mature leaves near the base portion of plant on main stem	20–30
Fruits and nuts			
Apple, apricot, almond, prune, peach, pear, cherry	Mid-season	Leaves near base of current year's growth or from spurs	50–100
Strawberry	Mid-season	Youngest fully expanded mature leaves	50–75
Pecan	6 to 8 weeks after bloom	Middle pair of leaflets from mid-portion of terminal growth	30–45
Walnut	6 to 8 weeks after bloom	Middle pair of leaflets from mature shoots	30–35
Lemon, lime	Mid-season	Mature leaves from last flush or growth on non-fruiting terminals	20–30
Orange	Mid-season	Spring cycle leaves, 4 to 7 months old from non-bearing terminals	20–30
Grapes	End of bloom period	Petioles from leaves adjacent to fruit clusters	60–100
Raspberry	Mid-season	Youngest mature leaves on lateral or "primo" canes	20–40

Suggested sampling procedures for field and vegetable crops, fruits and nuts, and ornamental plants (continued)

Crop	Stage of growth	Plant part to sample	Number of plants to sample
Ornamentals and flowers			
Ornamental trees, shrubs	Current year's growth	Fully developed leaves	30–100
Turf	During normal growing	Leaf blades; clip by hand to avoid contamination with soil or other material	1/4 liter
Roses	During flowering	Upper leaves on the flowering production stem	20–30
Chrysanthemums	Prior to or at flowering	Upper leaves on flowering stem	20–30
Carnations	Unpinched plants	Fourth or fifth leaf pairs from base of plant	20–30
	Pinched plants	Fifth and sixth leaf pairs from top of primary laterals	20–30
Poinsettias	Prior to or at flowering	Most recently mature fully expanded leaves	15–20

[a] Sampling after pods begin to set not recommended.
[b] Sampling after heading is not recommended.

If the sampling procedure used does not conform to that recommended, an interpretation of the plant analysis result may be difficult, if not impossible to make. Since there is a substantially large potential for error to occur due to improper sampling technique, only thoroughly trained and experienced technicians should be responsible for collecting tissue samples in the field.

The number of plants to sample in a particular situation depends on the general condition of the plants, soil homogeneity, and the purpose for which the analysis result will be used. To ensure representation, sampling as many plants as practical is recommended, collecting samples during a particular time of day and under calm climatic conditions.

THE GENERAL RULE

As a general rule, mature leaves exposed to full sunlight just below the growing tip on main branches or stems are usually preferred, taken just prior to or at the time the plants begin their reproductive stage of growth. In some situations, sampling may be necessary at earlier periods in the plant's growth cycle with the same maturity.

WHAT NOT TO SAMPLE

There are as many instructions on what *not* to sample as there are on what to sample. Plant tissue or plants not to sample are:

- Tissue covered with soil, dust, or residue chemicals.
- Plants damaged by insects, mechanically injured, or diseased.
- Tissue from dead plants or dead tissue.
- Plants under moisture or temperature stress.
- Plants markedly affected by nutritional stress.
- Border-row plants or end-row plants.
- Plants in weed-infested areas.
- Whole plants unless seedlings.

Whole young plants or plants beyond full maturity should not constitute the sample or a portion of the sample. Seeds are not normally useful for assessing the nutrient element status of plants, except possibly for the element N (Pierre et al., 1977a,b). In some instances, seed analyses have been of value in determining the molybdenum (Mo) and Zn supply for young plants developing from that seed (Reisenauer, 1956; Shaw et al., 1954).

Plants under stress due to a possible elemental deficiency or imbalance should be sampled only at the initiation of the stress. After a long (greater than 5 days) period of stress, plants develop unusual element concentrations in their tissues that can lead to a misinterpretation of a plant analysis result.

Normally, after pollination and as plants begin setting and developing fruit or seed, the elemental content of the vegetative portions of the plant begin to change substantially, making a plant analysis interpretation difficult if leaf tissue is collected at this time. Therefore, sampling after pollination is not recommended for most grain and fruit crops.

CROP LOGGING (TRACKING)/MONITORING

A major role for a plant analysis is crop logging (tracking) or monitoring, the taking of a series of plant tissue samples over the growing season. Crop logging (tracking) is more commonly used with plantation crops, such as sugarcane

(Clements, 1960; Bowen, 1990), and oil and date palm, while the monitoring of the nitrate-nitrogen (NO_3-N) and K petiole content of cotton (Sabbe and Zelinski, 1990; Constable et al., 1991; Davis, 1995) is widely used to regulate N fertilizer use in order to avoid either a deficiency or excess.

Maintaining a track of plant analysis results by element from season to season for the same crop is also useful in order to determine if there exists a changing trend in concentration, suggesting the development of a possible future insufficiency. In order for the track to be meaningful, care must be taken each year to ensure that samples collected for analysis represent the same physiological stage of development and plant part. The objective of tracking is to warn the farmer of corrective action needed before an insufficiency occurs.

COMPARATIVE SAMPLING

When visual symptoms occur, or a deficiency is suspected, the analysis of the same plant part from adjoining normal plants in the same field or area can aid in the interpretation (Munson and Nelson, 1990). However, if the plants being compared differ in their vigor and stage of development, the same plant part at the same stage of development may not exist. Therefore, comparing the analysis results between two sets of tissues may not be helpful to the interpretation.

It is advisable to collect soil samples from the same area where plants have been selected for sampling. Comparing soil and plant analysis results can greatly assist in the interpretation.

HANDLING PROCEDURES

Collected plant tissue is very perishable, requiring special handling to ensure that no loss in dry weight occurs as decomposition will reduce the dry weight, which in turn will significantly affect the plant analysis result (Lockman, 1970). Therefore, fresh plant tissue should be placed in open, clean paper bags, partially air-dried if possible, or kept in a cool environment during shipment to the laboratory. Fresh plant tissue should not be placed in closed plastic bags unless the tissue is either air-dried or the bag and contents are kept cool [40°F (4.4°C)].

For air-drying fresh plant tissue, place the tissue in an open, dry environment for 12 to 24 hours, a procedure that will remove much of the water in the tissue.

If the plant tissue collected is coated with soil, dust, and/or chemical residues that must be removed, decontamination must be done on the fresh tissue shortly after it has been collected. The procedure for decontaminating plant tissue is given in Chapter 3.

SUMMARY

Carefully following the recommended sampling technique cannot be overemphasized, since criteria for elemental analysis interpretation have been established for specific plant sampling procedures. Therefore, for elemental concentration determinations to be meaningful, it is essential to adhere to the given sampling procedures designed for that plant species and the element(s) to be assayed.

REFERENCES

Bates, T.E. 1971. Factors affecting critical nutrient concentrations in plants and their evaluation: A review. *Soil Sci.* 112:116–130.

Chapman, H.D. (Ed.). 1966. *Diagnostic Criteria for Plants and Soils.* Division of Agricultural Science, University of California, Riverside.

Clements, H.F. 1960. Crop logging of sugar cane in Hawaii, pp. 131–147. In: W. Reuther (Ed.). *Plant Analysis and Fertilizer Problems.* American Institute of Biological Sciences. Washington, D.C.

Colonna, J.P. 1970. The mineral diet of excelsior coffee plants. Natural variability of the mineral foliar composition on a homogeneous plantation. *Cashiers Office Res. Sci. Tech. Outre-Mer Ser. Biol.* 13:67–80.

Constable, G.A., I.J. Rochester, J.H. Betts, and D.F. Herridge. 1991. Prediction of nitrogen fertilizer requirement in cotton using petiole and sap nitrate. *Commun. Soil Sci. Plant Anal.* 22:1315–1324.

Davis, J.G. 1995. Impact of time of day and time since irrigation on cotton leaf blade and petiole nutrient concentration. *Commun. Soil Sci. Plant Anal.* 26:2351–2360.

Goodall, D.W. and F.G. Gregory. 1947. *Chemical Composition of Plants as an Index of their Nutritional Status.* Imperial Bureau Horticultural Plant Crops (GB) Technical Communication No. 17. Ministry of Agriculture, London, England.

Jones, Jr., J.B. 1963. Effect of drying on ion accumulation in corn leaf margins. *Agron. J.* 55:579–580.

Jones, Jr., J.B. 1967. Interpretation of plant analysis for several agronomic crops, pp. 49–58. In: G.W. Hardy (Ed.), *Soil Testing and Plant Analysis.* Plant Analysis. Part II. SSSA Special Publication No. 2. Soil Science Society of America, Madison, WI.

Jones, Jr., J.B. 1970. Distribution of 15 elements in corn leaves. *Commun. Soil Sci. Plant Anal.* 1:27–34.

Jones, Jr., J.B., R.L. Large, D.P. Pfeiderer, and H.S. Klosky. 1971. How to properly sample for a plant analysis. *Crops Soils* 23:15–18.

Jones, Jr., J.B., B. Wolf, and H.A. Mills. 1993. *Plant Analysis Handbook: A Practical Sampling, Preparation, Analysis, and Interpretation Guide.* Micro-Macro Publishing, Athens, GA.

Kenworthy, A.L. 1969. Fruit, Nut, and Plantation Crops Deciduous and Evergreen: A Guide for Collecting Foliar Samples for Nutrient Element Analysis. Horticultural Report No. 11. Michigan State University, East Lansing.

Lilleland, O. and J.G. Brown. 1943. Phosphate nutrition of fruit trees. *Proc. Am. Soc. Hort. Sci.* 41:1–10.

Lockman, R.B. 1970. Plant sample analysis as affected by sample decomposition prior to laboratory processing. *Commun. Soil Sci. Plant Anal.* 1:13–19.

Mills, H.A. and J.B. Jones, Jr. 1996. *Plant Nutrition Manual II.* Micro-Macro Publishing, Athens, GA.

Munson, R.D. and W.L. Nelson. 1990. Principles and practices in plant analysis, pp. 359–387. In: R.L. Westerman (Ed.), *Soil Testing and Plant Analysis.* SSSA Book Series 3. Soil Science Society of America, Madison, WI.

Pierre, W.H., V.D. Jolley, J.R. Webb, and W.D. Schrader. 1977a. Relation between corn yield, expressed as percentage of the maximum, and the N percentage in the grain. I. Various N-rate experiments. *Agron. J.* 69:215–220.

Pierre, W.H., L. Dumenil, and J. Henao. 1977b. Relationship between corn yield, expressed as percentage of the maximum, and the N percentage of the grain. II. Diagnostic use. *Agron. J.* 69:221–226.

Reisenauer, H.M. 1956. Molybdenum content of alfalfa in relation to deficiency symptoms and response to molybdenum fertilization. *Soil Sci.* 81:237–242.

Reuter, D.J. and J.B. Robinson (Eds.). 1986. *Plant Analysis: An Interpretation Manual.* Inkata Press Proprietary, Melbourne, Australia.

Reuter, D.J. and J.B. Robinson (Eds.). 1997. *Plant Analysis: An Interpretation Manual* (2nd Edition). CSIRO Publishing, Collingwood, Australia.

Sabbe, W.E. and L.J. Zelinski. 1990. Plant analysis as an aid in fertilizing cotton, pp. 469–493. In: R.L. Westerman (Ed.), *Soil Testing and Plant Analysis.* Third edition. Soil Science Society of America, Madison, WI.

Shaw, E., R.G. Menzel, and L.A. Dean. 1954. Plant uptake of zinc[65] from soils and fertilizers in the greenhouse. *Soil Sci.* 77:205–214.

Steyn, W.J.A. 1959. Leaf analysis. Errors involved in the preparative phase. *J. Agric. Food Chem.* 7:344–348.

Steyn, W.J.A. 1961. The errors involved in the sampling of citrus and pineapple plants for leaf analysis purposes, pp. 409–430. In: W. Reuther (Ed.), *Plant Analysis and Fertilizer Problems.* American Institute of Biological Science, Washington, D.C.

Thomas, W. 1945. Foliar diagnosis. *Soil Sci.* 59:353–374.

PREPARATION OF PLANT TISSUE FOR LABORATORY ANALYSIS

3

C. Ray Campbell and C. Owen Plank

INTRODUCTION

Sample preparation is critical in obtaining accurate analytical data and reliable interpretation of plant analysis results. Proven procedures should be followed during decontamination, drying, particle-size reduction, storage, and organic matter destruction. Each of these preparation procedures provides opportunities to enhance the accuracy and reliability of the analytical results.

DECONTAMINATION

Principle

Plant materials must be clean and free of extraneous substances, including soil and dust particles and foliar spray residues, that may influence analytical results. The elements most often affected by soil and dust particles are iron (Fe), aluminum (Al), silicon (Si), and manganese (Mn), especially with seedling and grass crops. Foliar nutrient spray and fungicide residues can affect several elements and should be taken into account in the decontamination process and when evaluating the analytical results. The decontamination process must be thorough while still preserving sample integrity. Therefore, decontamination

1-57444-124-8/98/$0.00+$.50
© 1998 by CRC Press LLC

procedures involving washing and rinsing should only be used for fresh, fully turgid plant samples.

Reagent and Apparatus

1. Deionized water.
2. 0.1 to 0.3% detergent solution (non-phosphate).
3. Medium-stiff nylon bristle brush or sponge.
4. Plastic containers suitable for washing and rinsing tissue samples.

Procedure

1. Examine fresh plant tissue samples to determine physical condition and extent of contamination. Unless leaf tissue is visibly coated with foreign substances, decontamination is usually not required except when Fe (Wallace et al., 1982; Jones and Wallace, 1992), Al, Si, or Mn are to be determined (Jones and Case, 1990).
2. When Al, Si, Mn, and Fe are not of primary interest, plant leaves should be brushed briskly to remove visible soil and dust particles.
3. When plant samples show visible residues from spray applications and when Al, Si, Fe and Mn are elements of interest, leaves should be washed in 0.1 to 0.3% detergent solution (Ashby, 1969; Wallace et al., 1982; Jones and Wallace, 1992), followed by rinsing in deionized water. The wash and rinse periods should be as short as possible (Sonneveld and van Dijk, 1982) to avoid danger of nitrate (NO_3), boron (B), potassium (K), and chloride (Cl) leaching from the tissue (Bhan et al., 1959; Smith and Storey, 1976).
4. After decontamination, samples should be dried immediately to stabilize the tissue and stop enzymatic reactions.

Comments

1. When proper sampling techniques have been utilized, decontamination should be minimized.
2. Decontamination is generally not necessary where tissue has been exposed to frequent rainfall and/or not exposed to nutrient or fungicidal sprays (Jones et al., 1991). Seedlings and pasture/turf crops that have been splattered with soil are the exception to this rule.
3. Excessive washing is likely worse than no decontamination since soluble elements including B, K, and NO_3-N are likely to leach from the tissue.
4. Samples should be dipped quickly in the wash and rinse solutions. Sonneveld and van Dijk (1982) recommended a time of 15 seconds.

5. Relatively high concentrations of Al (>100 mg kg^{-1}), Fe (>100 mg kg^{-1}), and Si ($>1.0\%$) are strong indicators of contamination (Jones et al., 1991). Titanium (Ti) has also been suggested as an indicator of soil or dust contamination (Cherney and Robinson, 1982).

OVEN DRYING

Principle

Water is removed from plant tissue to stop enzymatic reactions and to stabilize the sample. Removal of combined water also facilitates complete particle size reduction, thorough homogenization, and accurate weighing.

Apparatus

1. Forced-air oven equivalent to Blue M™, Model POM-166E.

Procedure

1. Separate or loosen tissue samples and place in paper containers.
2. Place container in forced-air oven and dry at 80°C for 12 to 24 hours. *Note:* The nature of the sample and its moisture content will affect the length of drying time. High carbohydrate-containing tissue may require another type of drying procedure.

Comments

1. Drying times longer than 24 hours may be required, depending on the type and number of plant samples in the dryer.
2. Drying at temperatures under 80°C may not remove all combined water (Jones et al., 1991) and may result in poor homogenization and incorrect analytical results.
3. Drying temperatures above 80°C may result in thermal decomposition and reduction in dry weight (Jones et al., 1991).
4. Enzymes present in plant tissue are rendered inactive at temperatures above 60°C (Tauber, 1949). As a result, air drying may not stabilize samples and prevent enzymatic decomposition. Samples should, therefore, be properly dried as soon after taking the sample as possible.
5. Quick drying of a limited number of samples can be accomplished using a microwave oven, provided the samples are turned often and the drying

process closely monitored (Carlier and van Hee, 1971; Shuman and Rauzi, 1981; Jones et al., 1991).

6. If samples absorb significant amounts of moisture during grinding, additional drying may be required prior to weighing for analysis. Drying time required will vary. Dry to constant weight by making periodic weighings.

PARTICLE SIZE REDUCTION

Principle

Plant tissue samples are reduced to 0.5 to 1.0-mm particle size to ensure homogeneity and to facilitate organic matter destruction.

Apparatus

1. Standard Wiley mill equipped with 20-, 40- and 60-mesh screens and stainless steel contact points, or a Cyclotec™ or equivalent high-speed grinder.
2. Medium bristle brush.
3. Vacuum system.

Procedure

1. After drying, samples should be ground to pass a 1.0-mm (20 mesh) screen using the appropriate Wiley™ mill. A 20-mesh sieve is adequate if the sample aliquot to be assayed is >0.5 g. However, if the sample aliquot to be assayed is <0.5 g, a 40-mesh screen should be utilized (Jones and Case, 1990).
2. After grinding, the sample should be thoroughly mixed and a 5- to 8-g aliquot withdrawn for analysis and storage.
3. Using a brush or vacuum system, clean the grinding apparatus after grinding each sample.   .

Comments

1. Uniform grinding and mixing are critical in obtaining accurate analytical results.
2. Exercise care when grinding very small samples or plant material that is pubescent, deliquescent, or that has a fibrous texture. These samples are difficult to grind in Wiley mills and the operator should allow sufficient time for the sample to pass through the screen to ensure homogeneity. In

these instances, experience has shown that Cyclotec or equivalent high-speed grinders are preferable.
3. Most mechanical mills contribute some contamination of the sample with one or more elements (Hood et al., 1944). The extent of contamination depends on the condition of the mill and exposure time (Jones and Case, 1990). Grier (1966) recommended use of stainless steel for cutting and sieving surfaces to minimize contamination.
4. Routine maintenance should be performed on mills to ensure optimum operating conditions. Cutting knives or blades should be maintained in sharp condition and in proper adjustment.

STORAGE

Principle

After particle size reduction and homogenization, samples should be stored in conditions that will minimize deterioration and maintain sample integrity for weighing and follow-up analytical work.

Apparatus

1. Air-tight plastic storage containers (5-dram size).
2. Storage cabinet located in cool, dark, moisture-free environment.
3. Refrigerator.

Procedure

1. After grinding and homogenization, a representative sample is taken from the ground plant material for analysis and storage. The sample should be placed in a container and securely sealed.
2. Containers should be stored under cool, dry conditions.
3. For long-term storage, ground samples should be thoroughly dried, sealed, and placed under refrigerated conditions (4°C) until analyses can be completed.

Comments

1. If samples are placed in a cool (4°C), dark, dry environment, storage life is indefinite (Jones et al., 1991).
2. Coin envelopes can also be used for sample storage; however, somewhat greater care must be exercised in sample handling to prevent absorption of

moisture. Collecting the ground sample in the envelope and immediately placing into a desiccation cabinet or jar will minimize moisture absorption.

ORGANIC MATTER DESTRUCTION

Plant tissue samples previously dried, ground, and weighed are prepared for elemental analysis through decomposition/destruction of organic matter. Extensive work has been done to evaluate published methods and to develop new and improved procedures. The best overviews on organic matter destruction are found in the books by Gorsuch (1970) and Bock (1978) and in the review articles by Tolg (1974) and Gorsuch (1976). The two commonly used methods of organic matter destruction are dry ashing (high-temperature combustion) and wet ashing (acid digestion) (Jones et al., 1991). Both methods are based on the oxidation of organic matter through the use of heat and/or acids. Detailed instructions are given in Chapters 5 through 8.

DRY ASHING

Principle

Dry ashing is conducted in a muffle furnace at temperatures of 500 to 550°C for 4 to 8 hours. For tissue high in carbohydrates and oils, ashing aids (Horwitz, 1980) may be required to achieve complete decomposition of organic matter. At the end of the ashing period, the vessel is removed from the muffle furnace, cooled, and the ash is dissolved in dilute nitric (HNO_3) or hydrochloric (HCl) acid, or a mixture of both, such as dilute *aqua regia*. The final solution is diluted as needed to meet the range requirements of the analytical procedure or instrument utilized.

Reagents and Apparatus

1. Muffle furnace with dual time and temperature control.
2. Fume hood.
3. Hot plate.
4. Porcelain or quartz crucibles, 30 mL.
5. Pyrex beakers, 50 mL.
6. Deionized water.
7. Hydrochloric acid (HCl), concentrated.
8. Nitric acid (HNO_3), concentrated.

9. Sulfuric acid (H_2SO_4), 10%.
10. Magnesium nitrate, [$Mg(NO_3)_2 \cdot H_2O$], 7%.
11. Dilute *aqua regia* (300 mL HCl and 100 mL HNO_3 in 1 L deionized water).

Procedure

1. Weigh 0.5 to 1.0 g dried (80°C) plant material that has been ground (0.5 to 1.0 mm) and homogenized into a high-form, 30-mL porcelain or quartz crucible or 100-mL Pyrex beaker.
2. Place samples in a cool muffle furnace.
3. Set temperature control of the furnace to allow a gradual increase (2 hours) in the ashing temperature (500 to 550°C) and maintain for 4 to 8 hours.
4. Turn furnace off, open door, and allow samples to cool.
5. Check the ash to determine extent of destruction. If a clean white ash is obtained, proceed with Step 9. If a clean white ash is not obtained, follow Step 1 with Step 6.
6. Moisten the tissue with concentrated HNO_3.
7. Place the container on a hot plate and evaporate the HNO_3 from the sample. Make sure the residue is completely free of moisture before placing into the muffle furnace.
8. Remove the container from the hot plate and repeat Steps 2, 3, and 4.
9. Depending on subsequent analytical procedures, the ash can be solubilized using the appropriate acid and/or mixture of acids.

Comments

1. Critical factors include selection of ashing vessel, sample number, placement in furnace, ashing temperature, time, selection of acid to solubilize the ash, and final volume (Jones and Case, 1990). The analyst has less latitude choosing an ashing temperature (Baker et al., 1964; Gorsuch, 1959, 1970, 1976; Isaac and Jones, 1972) than in selecting the other parameters of the digestion procedure. Placement of vessels and ashing time are dependent on the type and number of samples. Selection of an ashing vessel, the solubilizing acid and temperature (Munter and Grande, 1981) and final volume are dependent on the elements of interest and subsequent analytical procedures. Combinations of these factors have been used successfully.
2. If a clean white ash is obtained after muffling (oxidation), ashing aids are not required.
3. Plant materials with high sugar or oil content (highly carbonaceous) may require an ashing aid. Aids commonly used are 10% H_2SO_4, concentrated

HNO_3, or 7% [$Mg(NO_3)_2 \cdot H_2O$] solutions. The latter is recommended when the tissue is to be assayed for sulfur (S) as sulfate ($SO_4^=$). Gorsuch (1970), Horwitz (1980), and Jones et al. (1991) provided details on the use of these oxidizing aids.

4. Dry ashing is not recommended for plant materials that are high in Si as low micronutrient recoveries, especially zinc (Zn), are frequently obtained.
5. Dry ashing techniques may result in lower Fe and Al values compared to wet ashing techniques (Jones and Case, 1990).

WET ASHING

Principle

Wet digestion involves the destruction of organic matter through the use of both heat and acids. Acids that have been used in these procedures include sulfuric (H_2SO_4), nitric acid (HNO_3), and perchloric ($HClO_4$) acids, either alone or in combination. Hydrogen peroxide (H_2O_2) is also used to enhance reaction speed and to complete the digestion. Most laboratories have eliminated the use of $HClO_4$ due to risk of explosion, as well as safety regulations that require specially designed hoods where $HClO_4$ is used. Hot plates or digestion blocks are frequently used to maintain temperatures of 80 to 125°C. After digestion is complete and the sample is cooled, dilutions are made to meet analytical requirements.

Reagents and Apparatus

1. Hot plate.
2. Block digester.
3. Fumehood (if $HClO_4$ is used, specific conditions required).
4. Nitric acid (HNO_3), concentrated.
5. Sulfuric acid (H_2SO_4), concentrated.
6. Hydrogen peroxide (H_2O_2), 30%.
7. Perchloric acid ($HClO_4$), 60%
8. 200-mL tall-form beakers or digestion tubes.
9. Deionized water.

Procedure

1. Weigh 0.5 to 1.0 g dried (80°C) plant material that has been ground (0.5 to 1.0 mm) and thoroughly mixed and place in a tall-form beaker or digestion tube.

2. Add 5.0 mL conc. HNO_3 and cover beaker with watch glass or place a funnel in the mouth of the digestion tube and allow to stand overnight or until frothing subsides.
3. Place covered beaker on hot plate or digestion tube into block digester and heat at 125°C for 1 hour (where elemental analysis is by inductively coupled plasma emission spectrometry (ICP-AES), the digestion time can be extended to 4 hours and Steps 5 and 6 omitted).
4. Remove beaker or digestion tube and allow to cool.
5. Add 1 to 2 mL 30% H_2O_2 and digest at the same temperature. Repeat heating and 30% H_2O_2 additions until digest is clear. Add additional HNO_3 as needed to maintain the digest volume.
6. After sample digest is clear, remove watch glass or funnel and lower temperature to 80°C. Continue heating until near dryness. The residue should be clear or white if digestion is complete.
7. Add dilute HNO_3, HCl or a combination of the two acids and deioinized water to dissolve digest residue and bring sample to final volume, depending on requirements of the subsequent analytical procedure to be used.
8. If $HClO_4$ is to be used, complete Steps 1 through 4. Add 2 mL HClO4 and return the beaker or digestion tube to the hot plate or block digester. Heat until white fumes are produced. If a beaker is the digestion vessel, reduce the hot plate temperature and continue heating until the remaining residue is just moist. Remove the beaker from the hot plate and let cool. The residue should be white or very light-yellow in color. If brown or dark yellow in color, add 1 mL $HClO_4$ and return to the hot plate. When the residue is white or very light-yellow in color, follow the procedure given in Step 7.

Comments

1. Critical factors in wet digestion procedures include selection of the digestion vessel, temperature and control, time, the digestion mixture, and final volume. Selection of a digestion vessel is dependent on the elements of interest and the heat source. Digestion blocks have been developed (Gallaher et al., 1975; Tucker, 1974) and used successfully. They shorten digestion time and allow very uniform temperature control. Time and temperature are interrelated and are dependent on the digestion mixture. A number of ashing mixtures have been recommended and include those reported by Parkenson and Allen (1975), Zasoski and Buran (1977), Cresser and Parsons (1979), Wolf (1982), Adler and Wilcox (1985), Huang and Schulte (1985), Zarcinas et al. (1987), Jones and Case (1990), and Jones et al. (1991). Wet digestion procedures generally require greater analyst supervision and intervention than the dry ash procedure.

2. Nitric acid is used in most wet oxidation procedures. The addition of H_2SO_4 is used to raise digestion temperature while $HClO_4$ or 30% H_2O_2 are used to increase speed of the reaction and ensure complete digestion (Jones and Case, 1990).
3. Most wet digestion procedures can be completed using covered beakers on hot plates but digestion blocks are preferred due to enhanced temperature control.
4. Wet ashing is recommended for plant materials that are high in Si or that contain volatile elements [arsenic (As), mercury (Hg), or selenium (Se)] that may be lost during the dry ash procedure.
5. Wet ashing techniques may result in higher Fe and Al values, compared to that obtained using the dry ashing technique (Jones and Case, 1990).

ACCELERATED WET DIGESTION

Principle

Relatively new alternatives for organic matter destruction include wet oxidation procedures, which utilize pressure and/or high temperature to shorten digestion time. Closed or open vessels are used either with conventional hot plates or in microwave ovens (White and Douthit, 1985).

A number of procedures have been developed that utilize microwave as a source of heat. These are generally classified as closed or open vessel. Closed vessel (Parr Bomb) utilizes heat and pressure to increase reaction rate and to decrease digestion time (Vigler et al., 1980; Okamoto and Fuwa, 1984). Element loss is controlled with a reflux valve. Open vessel procedures do not utilize pressure containers and must be monitored closely to avoid excess frothing and sample loss. The following procedure was developed by Campbell and Whitfield (1991) for the digestion of a wide variety of plant samples.

Reagents and Apparatus

1. CEM Microwave Digestion System (CEM Corporation, P.O. Box 9, Indian Trail, NC 28079).
2. Fumehood and scrubber.
3. Nitric acid (HNO_3), concentrated.
4. Hydrogen peroxide (H_2O_2), 30%.
5. 50-mL Erlenmeyer flask (must be heat-acid washed and relatively free from scratches to avoid B contamination).
6. Deionized water.

Procedure

1. Transfer 0.5 to 1.0 g dried (80°C) plant material that has been ground to 1.0 mm and thoroughly homogenized into a 50-mL Erlenmeyer flask.
2. Add 10 to 15 mL (10 mL for a 0.5 g sample) conc. HNO_3 to each sample and swirl the flask gently so that all the plant material comes in contact with the acid.
3. Place in specially designed microwave oven (see Comments) and digest for 30 minutes at 30% power (210 watts).
4. Flush sides of Erlenmeyer flasks with 30% H_2O_2.
5. Digest for 5 minutes at 60% power (390 watts).
6. While sample is still warm, fill to 50-mL volume with deionized water and shake well.
7. Filter digest (Whatman No. 1), and transfer a 10-mL aliquot into suitable container for analysis.
8. Digest is ready for analysis with or without further dilution. The procedure is designed for the elemental analysis to be done by ICP- AES.

Comments

Microwave digestion procedures require the use of specially designed ovens to handle acid fumes. Ideally, the microwave exhaust should be passed through a scrubber before being released into the fume exhaust system. Special safety precautions are required for microwave digestion (see manufacturer's specifications for details).

REFERENCES

Adler, P.R. and G.E. Wilcox. 1985. Rapid perchloric acid methods for analysis of major elements in plant tissue. *Commun. Soil Sci. Plant Anal.* 16:1153–1163.

Ashby, D.L. 1969. Washing techniques for the removal of nutrient element deposits from the surface of apple, cherry, and peach leaves. *J. Am. Soc. Hort. Sci.* 94:266–268.

Bahn, K.C., A. Wallace, and D.R. Hunt. 1959. Some mineral losses from leaves by leaching. *Proc. Am. Soc. Hort. Sci.* 73:289–293.

Baker, J.H., G.W. Gorsline, C.G. Smith, W.I. Thomas, W.E. Grube, and J.L. Ragland. 1964. Technique for rapid analyses of corn leaves for eleven elements. *Agron. J.* 56:133–136.

Bock, R.A. 1978. *Handbook of Decomposition Methods in Analytical Chemistry.* International Textbook Co., Glasgow, Scotland.

Campbell, C.R. and K. Whitfield. 1991. A rapid open vessel procedure for microwave digestion of plant tissue for analysis by ICP emission spectroscopy. Personal communication.

Carlier, L.A. and L.P. van Hee. 1971. Microwave drying of lucerne and grass samples. *J. Sci. Food Agric.* 22:306–307.

Cherney, J.H. and D.L. Robinson. 1982. A comparison of plant digestion methods for identifying contamination of plant tissue by Ti analysis. *Agron. J.* 75:145–147.

Cresser, M.S. and J.W. Parsons. 1979. Sulfuric-perchloric acid digestion of plant material for the determination of nitrogen, phosphorus, potassium, calcium, and magnesium. *Anal. Chem. Acta* 109:431–436.

Gallaher, R.N., C.O. Wilson, and J.G. Futral. 1975. An aluminum block digester for plant and soil analysis. *Soil Sci. Soc. Am. Proc.* 39:803–806.

Gorsuch, T.T. 1959. Radiochemical investigation on the recovery for analysis of trace elements in organic and biological materials. *Analyst* 84:135–173.

Gorsuch, T.T. 1970. *Destruction of Organic Matter.* International Series of Monographs in Analytical Chemistry. Volume 39. Pergamon Press, New York.

Gorsuch, T.T. 1976. Dissolution of organic matter, pp. 491–508. In: P.D. Lafluer (Ed.), *Accuracy in Trace Analysis. Sampling, Sample Handling, Analysis.* Volume 1. Special Publication 422. National Bureau of Standards, Washington, D.C.

Grier, J.D. 1966. Preparation of plant material for plant analysis. *J. Assoc. Off. Anal. Chem.* 49:292–298.

Hood, S.L., R.Q. Parks, and C. Hurwitz. 1944. Mineral contamination resulting from grinding plant samples. *Ind. Eng. Chem. Anal. Ed.* 16:202–205.

Horwitz, W. (Ed.). 1980. *Official Methods of Analysis of the Association of Official Analytical Chemists.* Thirteenth edition. Association of Official Analytical Chemists, Arlington, VA.

Huang, C.L. and E.E. Schulte. 1985. Digestion of plant tissue for analysis by ICP emission spectroscopy. *Commun. Soil Sci. Plant Anal.* 16:943–958.

Isaac, R.A. and J.B. Jones, Jr. 1972. Effects of various drying temperatures on the determination of five plant tissues. Commun. Soil Sci. Plant Anal. 3:261-269.

Jones, Jr., J.B. and V.W. Case. 1990. Sampling, handling, and analyzing plant tissue samples, pp. 389–427. In: R.L. Westerman (Ed.), *Soil Testing and Plant Analysis.* SSSA Book Series 3. Soil Science Society of America, Madison, WI.

Jones, Jr., J.B., B. Wolf, and H.A. Mills. 1991. *Plant Analysis Handbook,* pp. 23–26. Micro- Macro Publishing, Athens, GA.

Jones, Jr., J.B. and A. Wallace. 1992. Sample preparation and determination of iron in plant tissue samples. *J. Plant Nutr.* 15:2085–2108.

Munter, R.C. and R.A. Grande. 1981. Plant analysis and soil extracts by ICP-atomic emission spectrometry, pp. 653–672. In: R.M. Barnes (Ed.), *Developments in Atomic Plasma Spectrochemical Analysis.* Heyden and Son, Ltd., London, England.

Okamoto, K. and K. Fuwa. 1984. Low-contamination digestion bomb method using a Teflon double vessel for biological materials. *Anal. Chem.* 56:1756–1760.

Parkenson, J.A. and S.E. Allen. 1975. A wet oxidation procedure suitable for the determination of nitrogen and mineral nutrients in biological material. *Commun. Soil Sci. Plant Anal.* 6:1–11.

Shuman, G.E. and F. Ruazi. 1981. Microwave drying of rangeland forage samples. *J. Range Manage.* 34:426–428.

Smith, M. and J.B. Storey. 1976. The influence of washing procedure on surface removal and leaching of certain elements from trees. *Hort. Sci.* 14:718–719.

Sonneveld, C. and P.A. van Dijk. 1982. The effectiveness of some washing procedures on the removal of contaminants from plant tissue of glasshouse crops. *Commun. Soil Sci. Plant Anal.* 13:487–496.

Tauber, H. 1949. *Chemistry and Technology of Enzymes.* John Wiley and Sons, New York.

Tolg, G. 1974. The basis of trace analysis, pp. 698–710. In: E. Korte (Ed.), *Methodium Chimicum.* Volume 1. Analytical Methods. Part B. Micromethods, Biological Methods, Quality Control, Automation. Academic Press, New York.

Tucker, M.R. 1974. A modified heating block for plant tissue digestion. *Commun. Soil Sci. Plant Anal.* 5:539–546.

Vigler, M.S., A.W. Varnes, and H.A. Strecker. 1980. Sample preparation techniques for AA and ICP spectroscopy. *Am. Lab.* 12:21–34.

Wallace, A., J. Kinnear, J.W. Cha, and E.M. Romney. 1982. Influence of washing of soybean leaves on identification of iron deficiency by leaf analysis. *J. Plant Nutr.* 5:805–810.

White, Jr., R.T. and G.E. Douthit. 1985. Use of microwave oven and nitric acid-hydrogen peroxide digestion to prepare botanical materials for elemental analysis by inductively coupled argon plasma emission spectroscopy. *J. Assoc. Off. Anal. Chem.* 68:766–769.

Wolf, B. 1982. A comprehensive system of leaf analysis and its use for diagnosing crop nutrient status. *Commun. Soil Sci. Plant Anal.* 13:1035–1059.

Zarcinas, B.A., B. Cartwright, and L.P. Spauncer. 1987. Nitric acid digestion and multi-element analysis plant material by inductively coupled plasma spectrometry. *Commun. Soil Sci. Plant Anal.* 18:131–147.

Zasoski, R.J. and R.G. Buran. 1977. A rapid nitric-perchloric acid digestion method for multielement tissue analysis. *Commun. Soil Sci. Plant Anal.* 3:425–436.

DETERMINATION OF DRY MATTER CONTENT OF PLANT TISSUE: GRAVIMETRIC MOISTURE

4

Robert O. Miller

SCOPE AND APPLICATION

This method quantitatively determines the dry-matter percentage in plant tissues based on the gravimetric loss of free water associated with heating to 105°C for a period of 2 hours. The dry-matter determination is used to correct the sample element concentration to an absolute dry-matter basis. The method does not remove molecular bound water and is generally reproducible within ±7%.

EQUIPMENT

1. Analytical balance, 250-g capacity, resolution ±1.0 mg.
2. Aluminum weight dish with handle.
3. Drying oven, preheated to 105°C.
4. Desiccator, containing a desiccating agent.

1-57444-124-8/98/$0.00+$.50
© 1998 by CRC Press LLC

PROCEDURE

1. Weigh approximately 2 g air-dry plant tissue into a tared aluminum weigh pan (preweighed to nearest 0.001 g) and record moist sample weight to the nearest 0.001 g.
2. Place sample and weigh pan in drying oven set at 105°C for a minimum of 2 hours.
3. Remove and place pan in desiccator for 1 hour.
4. Weigh sample and pan on balance to the nearest 0.001 g.
5. Dispose of sample (see Comment).

CALCULATION

$$\text{Sample dry matter \%} = \frac{[1\text{-(sample moist wt.)} - \text{(sample dry wt.} - \text{pan tare wt.)}]}{\text{(sample dry weight pan tared weight)}} \times 100$$

Report dry matter content to the nearest 0.1%.

COMMENT

Drying samples at 105°C may volatilize some carbon, nitrogen, and sulfur compounds. Therefore, material used for moisture content should not be used for inorganic analysis.

REFERENCES

Reuter, D.J., J.B. Robinson, K.I. Peverill, and G.H. Price. 1986. Guidelines for collecting, handling, and analyzing plant materials, pp. 11–35. In: D.J. Reuter and J.B. Robinson (Eds.), *Plant Analysis: An Interpretation Manual.* Inkata Press, Melbourne, Australia.

Smith, M. and J. B. Storey. 1976. The influence of washing procedures on surface removal and leaching of certain elements from trees. *HortSci.* 14:718–719.

HIGH-TEMPERATURE OXIDATION: DRY ASHING

5

Robert O. Miller

SCOPE AND APPLICATION

This method prepares plant tissue for the quantitative determination of the concentration of boron (B), calcium (Ca), copper (Cu), iron (Fe), magnesium (Mg), manganese (Mn), phosphorus (P), potassium (K), sodium (Na), and zinc (Zn) utilizing high-temperature dry oxidation of the organic matter (Baker et al., 1964) and dissolution of the ash with hydrochloric acid (HCl). The best reviews on organic matter destruction are books by Gorsuch (1970) and Bock (1978) and the review articles by Gorsuch (1976) and Tolg (1974).

Digest analyte concentrations can be determined by either atomic absorption spectrometry (AAS) (Watson and Isaac, 1990) and/or inductively coupled plasma atomic emission spectrometry (ICP-AES) (Munter and Grande, 1981). Analysis of B and P may be conducted using spectrophotometric methods (Chapman and Pratt, 1961; Gaines and Mitchell, 1979; Loshe, 1982). The procedure is not quantitative for sulfur (S) and other analytes [i.e., arsenic (As), mercury (Hg), and selenium (Se)], which are easily volatilized. Ashing temperatures exceeding 500°C will result in poor recoveries of Al, B, Cu, Fe, K, and Mn (Isaac and Jones, 1972; Labanauskas and Handy, 1973). Results for B may be inconsistent due to volatilization and deabsorption in the muffle furnace. The method detection limit is approximately 0.04% for Ca, K, Mg, and

1-57444-124-8/98/$0.00+$.50
© 1998 by CRC Press LLC

P and 4.0 mg/kg for Cu, Fe, Mn, Na, and Zn on a dry-sample basis. The method is generally reproducible within ±7%.

EQUIPMENT

1. Analytical balance, 250-g capacity, resolution ±1.0 mg.
2. Porcelain crucibles, 30-mL capacity.
3. Muffle furnace capable of 500°C.
4. Repipette, 10.0 ± 0.2 mL.
5. Volumetric labware, 50 mL, plastic.
6. AAS and/or ICP-AES.

REAGENTS

1. Deionized water, ASTM Type I grade.
2. **1.0N HCl Solution:** dilute 83.5 mL concentrated HCl to 1.0 L with deionized water.
3. **Standard Calibration Solutions** (P, K, Ca, Mg, Na, Zn, Mn, Cu, and Fe): from 1,000 mg/L reference solutions, prepare five multielement standards of K, Ca, and Mg ranging from 5.0 to 500 mg/L; P and Na ranging from 1.0 to 100 mg/L; and Na, Zn, Mn, Fe, and Cu ranging from 0.10 to 10.0 mg/L. Dilute standard calibration solutions with 0.1N HCl.

PROCEDURE

1. Weigh 500 ± 5.0 mg plant material into a porcelain crucible. Include a method blank (see Comments 1 and 2).
2. Place crucible in a muffle furnace and slowly increase the ramp temperature to 500°C over 2 hours. Ash samples for 4 hours at 500°C (see Comment 3).
3. Allow to cool to room temperature in muffle furnace, slowly open door, and remove ashed samples. Take caution not to disturb sample ash while transferring from the furnace.
4. Dissolve ash with 10.0 mL 1.0N HCl solution (see Comments 4 and 5). Dissolution of the ash and recovery of some elements may be facilitated by heating (Munter and Grande, 1981).
5. Quantitatively transfer the contents of the crucible into a 50-mL volumetric flask, dilute to volume with deionized water, cap, and invert three times.

6. Elemental analysis of plant digests can be made using AES, AAS, ICP-AES, and/or other methodologies (see Comments 6 and 7). The method chosen will determine specific matrix modifications, calibration standard range, and the need for instrument-specific sample preparations and dilutions. Adjust and operate instruments in accordance with manufacturer's instructions. Calibrate instrument using standard calibration solutions. Determine the analyte concentrations of a method blank, unknown samples and record analyte concentrations in mg/L.

CALCULATIONS

For P, K, Ca, Mg, and Na, report results to the nearest 0.001%:

$$\% \text{ analyte} = \frac{(\text{mg/L} - \text{method blank}) \times (25) \times (0.0001)}{\text{dry matter (\%)}/100}$$

For B, Zn, Mn, and Fe, report results to the nearest 1 mg/kg, and for Cu, the nearest 0.1 mg/kg:

$$\text{mg/kg analyte} = \frac{(\text{mg/L} - \text{method blank}) \times (25)}{\text{dry matter (\%)}/100}$$

COMMENTS

1. Labware cleaning: (1) soak crucibles in 1% solution of labware detergent for 1 hour; (2) rinse vessels in tap water; (3) rinse in solution of 0.5N HCl; (4) three deionized water rinses (ASTM Type I grade); and (5) dry for 1 hour at 80°C.
2. Sample material must be ground to pass a 40-mesh screen to ensure homogeneity.
3. Ashing temperatures are not to exceed 500°C to avoid potential volatilization of Al, B, Cu, K, and Mn.
4. Check pipette dispensing volume, calibrate using an analytical balance.
5. When adding reagent to vessels and handling digests always wear protective clothing (i.e., eye protection, lab coat, and disposable gloves and shoes). Always handle reagents and opening of vessels in an acid hood capable of high air flow at 100 cfm.
6. Centrifuging may be necessary to clear the digest.

7. Samples having analyte concentrations exceeding the highest standard will require dilution and reanalysis.
8. This technique is used for the determination of P, K, Ca, Mg, Na, B, Cu, Fe, Mn, and Zn.

REFERENCES

Baker, D.E., G.W. Gorsline, C.G. Smith, W.I. Thomas, W.E. Grube, and J.L. Ragland. 1964. Techniques for rapid analysis of corn leaves for eleven elements. *Agron. J.* 56:133–136.

Bock, R.A. 1978. *Handbook of Decomposition Methods in Analytical Chemistry.* International Textbook Co., Glasgow, Scotland.

Chapman, H.D. and P.F. Pratt. 1961. Methods of Analysis for Soils, Plants and Waters. Priced Publication 4034. University of California–Berkeley, Division of Agricultural Science.

Gaines, T.P. and G.A. Mitchell. 1979. Boron determination in plant tissues by the Azomethine-H method. *Commun. Soil Sci. Plant Anal.* 10:1099–1108.

Gorsuch, T.T. 1970. *Destruction of Organic Matter.* International Series of Monographs in Analytical Chemistry. Volume 39. Pergamon Press, New York.

Gorsuch, T.T. 1976. Dissolution of organic matter, pp. 491–508. In: P.D. LaFluer (Ed.), *Accuracy in Trace Analysis. Sampling, Sample Handling, Analysis.* Volume 1. Special Publication 422. National Bureau of Standards, Washington, D.C.

Isaac, R.A. and J.B. Jones, Jr. 1972. Effects of various dry ashing temperatures on the determination of 13 elements in five plant tissues. *Commun. Soil Sci. Plant Anal.* 3:261–269.

Labanauskas, C.K. and M.F. Handy. 1973. Time, temperature and sample size effects in dry ashing of plant tissue. IN: *Summaries, Advances and Innovations in Progress for Soil, Plant, and Water Analysis.* University of California, Berkeley.

Loshe, G. 1982. Microanalytical Azomethine-H method for boron determination in plant tissue. *Commun. Soil Sci. Plant Anal.* 13:127–134.

Munter, R.C. and R.A. Grande. 1981. Plant analysis and soil extracts by ICP-atomic emission spectrometry, pp. 653–672. IN: R.M. Barnes (Ed.), *Developments in Atomic Plasma Spectrochemical Analysis.* Heyden and Son, Ltd., London, England.

Tolg, G. 1974. The basis of trace analysis, pp. 698–710. In: E. Korte (Ed.), *Methodium Chimicuin.* Volume 1. Analytical Methods. Part B. Micromethods, Biological Methods, Quality Control, Automation. Academic Press, New York.

Watson, M.E. and R.A. Isaac. 1990. Analytical instruments for soil and plant analysis, pp. 691–740. In: R.L. Westerman (Ed.), *Soil Testing and Plant Analysis.* Third edition. SSSA Book Series 3. Soil Science Society of America, Madison, WI.

NITRIC-PERCHLORIC ACID WET DIGESTION IN AN OPEN VESSEL

6

Robert O. Miller

SCOPE AND APPLICATION

The method prepares plant tissue for the quantitative determination of the concentration of calcium (Ca), copper (Cu), iron (Fe), magnesium (Mg), manganese (Mn), molybdenum (Mo), phosphorus (P), potassium (K), selenium (Se), sodium (Na), sulfur (S), and zinc (Zn), and trace elements using a nitric-perchloric (HNO_3-$HClO_4$) acid digestion of organic matter in conjunction with external heating (Gorsuch, 1970, 1976; Tolg; 1974; Zasoski and Burau, 1977; Bock, 1978; Jones and Case, 1990). Digest analyte concentrations can be determined by either atomic absorption spectrometry (AAS) (Watson and Isaac, 1990) and/or inductively coupled plasma atomic emission spectrometry (ICP-AES) (Munter and Grande, 1981). Analysis of P and S may be conducted using spectrophotometric and turbidimetric methods, respectively (Johnson and Ulrich, 1959; Chapman and Pratt, 1961; Blanchar, 1986). The method requires predigestion with HNO_3, followed by addition of $HClO_4$ and digestion at high temperatures. Extreme caution is to be exercised when using $HClO_4$ (Schilt, 1979), which may react violently with untreated organic materials and result in an explosion. A special hood is required to handle $HClO_4$ fumes. Reflux funnels are placed over the digestion tubes to reduce volatilization and minimize oxygen (O_2) infusion. Alternatives to the use of $HClO_4$ have been reported (Haung

1-57444-124-8/98/$0.00+$.50
© 1998 by CRC Press LLC

and Schulte, 1985; Havlin and Soltanpour, 1980); however, there may be a question of quantitative recoveries. The method can also be used for the determination of trace elements [lead (Pb), nickel (Ni), cadmium (Cd), etc.]. Boron is not included, since digestion labware often contains B materials and B can be volatilized from some types of plant tissue during the digestion process. Generally, the method detection limit is approximately 0.02% for P, S, K, Ca, Mg, and Na; and 0.5 mg/kg (sample dry basis) for Zn, Mn, Fe, and Cu. Generally, reproducibility is within ±7.0%.

EQUIPMENT

1. Analytical balance, 250-g capacity, resolution ±0.1 mg.
2. Block digester (400°C) and perchloric acid fume hood.
3. 50-mL volumetric digestion tubes and 25-mm reflux funnels.
4. Repipette dispensers, calibrated to 6.0 ± 0.05 mL and 2.0 ± 0.01 mL.
5. Volumetric labware, 25 mL.
6. AAS and/or ICP-AES(vacuum or purged system).

REAGENTS

1. Deionized water, ASTM Type 1 grade.
2. Nitric acid (HNO_3), concentrated, reagent grade.
3. Perchloric acid ($HClO_4$), 70%, reagent grade.
4. Standard calibration solutions (P, K, S, Ca, Mg, Na, Zn, Mn, Cu, and Fe), 1,000 mg/L.
   Prepare five multielement standards:
   • K, Ca, Mg ranging from 5 to 500 mg/L
   • P, S, and Na ranging from 1.0 to 100 mg/L
   • Zn, Mn, Fe, and Cu ranging from 0.10 to 10.0 mg/L
   Dilute standard calibration solutions with 5% HNO_3 and 1% $HClO_4$ by volume.

PROCEDURE

1. Weigh 500.0 ± 0.5 mg of the sample into a 50-mL volumetric digestion tube (see Comments 1 and 2). Include a method blank.

2. Using a repipette, add 6.0 mL HNO_3, a boiling chip (Teflon or glass) and swirl to thoroughly wet the sample (see Comments 3, 4, and 5).
3. Place 25-mm reflux funnels over the samples and allow to predigest at room temperature for 60 minutes (overnight preferred).
4. Place the digestion tubes into a digestion block port for 60 minutes at 150°C.
5. Remove, cool to room temperature and, using a repipette, slowly add 2.0 mL $HClO_4$ through the funnels.
6. Place the tubes back into a digestion block port at a block temperature of 215°C for 2 hours after the HNO_3 fumes have evolved.
7. Remove the funnels 10 minutes before the end of the digestion.
8. Remove the tubes from the digestion block, cool 20 minutes in a hood, and add 10 mL deionized water on a hot plate (90°C).
9. Mix, using a vortex stirrer, cool, and dilute to final volume. Filtering or centrifuging may be necessary to remove all particulate matter in the digest prior to analysis. Quantitatively transfer contents of digestion tube into a 25-mL volumetric flask.
10. Elemental analysis of plant digests can be made using AES, AAS, ICP-AES, and/or other analytical methods. The method chosen will determine specific matrix modifications and the need for instrument-specific sample preparations and dilutions. Adjust and operate instruments in accordance with manufacturer's instructions. Calibrate instrument using calibration solutions and record concentration of analytes as mg/L (see Comments 6, 7, and 8). Determine the analyte concentrations of a method blank, unknown samples and record concentrations in mg/L.

CALCULATIONS

Report P, K, S, Ca, Mg, and Na results to the nearest 0.001%:

$$\% \text{ analyte} = \frac{(\text{mg/L} - \text{method blank}) \times (50) \times (0.0001)}{\text{dry matter } (\%)/100}$$

Report Zn, Mn, and Fe results to the nearest 1 mg/kg, and Cu to the nearest 0.1 mg/kg:

$$\text{mg/kg analyte} = \frac{(\text{mg/L} - \text{method blank}) \times (50)}{\text{dry matter } (\%)/100}$$

COMMENTS

1. Labware cleaning: (1) soak digestion tubes in 1% solution of labware detergent for 1 hour, (2) rinse vessels in tap water, (3) rinse in solution of $0.5N$ HCl, (4) three deionized water rinses (ASTM Type I grade), and (5) dry for 1 hour at 80°C.
2. Sample material must be ground to pass a 40-mesh screen (<400 mM opening) to ensure homogeneity.
3. When adding reagent to vessels and handling digests, always wear protective clothing (i.e., eye protection, lab coat, disposable gloves and shoes). Always handle reagents and opening of vessels in an acid hood capable of high air flow at 100 cfm.
4. It is essential that the entire sample be pretreated with HNO_3 to ensure at least partial oxidation of the organic matter before the addition of $HClO_4$. *Caution:* $HClO_4$ in the presence of untreated organic matter can lead to rapid oxidation of the sample and a possible explosion (Schilt, 1979).
5. Check repipette dispensing volume, calibrate using an analytical balance.
6. The method may not be quantitative for K, since this alkali metal may form the precipitate, potassium perchlorate.
7. Samples having analyte concentrations exceeding the highest standard will require dilution and reanalysis.
8. Increase sample mass for the determination of trace metals, such as Cd, chromium, and barium to 1,000 mg.
9. This technique is for the determination of the elements, P, K, Ca, Mg, S, Cu, Fe, Mn, and Zn.

REFERENCES

Blanchar, R.W. 1986. Measurements of sulfur in soils and plants. In: M.A. Tabatabai (Ed.), *Sulfur in Agriculture.* American Society of Agronomy, Madison, WI.

Bock, R. 1978. *Handbook of Decomposition Methods in Analytical Chemistry.* International Textbook Co., Glasgow, Scotland.

Chapman, H.D. and P.F. Pratt. 1961. Methods of Analysis for Soils, Plants, and Waters. Price Publication 4034. University of California–Berkeley, Division of Agriculture.

Gorsuch, T.T. 1970. *Destruction of Organic Matter.* International Series of Monographs in Analytical Chemistry. Volume 39. Pergamon Press, New York.

Gorsuch, T.T. 1976. Dissolution of organic matter, pp. 491–508. In: P.D. LaFluer (Ed.), *Accuracy in Trace Analysis. Sampling, Sample Handling, Analysis.* Volume 1. Special Publication 422. National Bureau of Standards, Washington, D.C.

Haung, C.L. and E.E. Schulte. 1985. Digestion of plant tissue for analysis by ICP emission spectroscopy. *Commun. Soil Sci. Plant Anal.* 16: 945–958.

Havlin, J.L. and P.N. Soltanpour. 1980. A nitric acid plant tissue digest method for use with inductively coupled plasma spectrometry. *Commun. Soil Sci. Plant Anal.* 11:969–980.

Johnson, C.M. and A. Ulrich. 1959. *Analytical Methods for Use in Plant Analysis,* pp. 26–78. Bulletin 766. University of California Experiment Station, Berkeley.

Jones, Jr., J.B. and V.W. Case. 1990. Sampling, handling, and analyzing plant tissue samples, pp. 389–427. In: R.L. Westerman (Ed.), *Soil Testing and Plant Analysis.* Third edition. SSSA Book Series 3. Soil Science Society of America, Madison, WI.

Munter, R.C. and R.A. Grande. 1981. Plant analysis and soil extracts by ICP-atomic emission spectrometry, pp. 653–672. In: R.M. Barnes (Ed.), *Developments in Atomic Plasma Spectrochemical Analysis.* Heyden and Son, Ltd., London, England.

Schilt, A.A. 1979. *Perchloric Acid and Perchlorates.* Frederick Smith Chemical Company, Columbus, OH.

Tolg, G. 1974. The basics of trace analysis, pp. 698–710. In: E. Korte (Ed.), *Methodium Chimicun.* Volume 1. Analytical Methods. Part 19. Micromethods, Biological Methods, Quality Control, Automation. Academic Press, New York.

Watson, M.E. and R.A. Isaac. 1990. Analytical instruments for soil and plant analysis, pp. 691–740. In: R.L. Westerman (Ed.), *Soil Testing and Plant Analysis.* Third edition. SSSA Book Series 3. Soil Science Society of America, Madison, WI.

Zasoski, R.J. and R.G. Burau. 1977. A rapid nitric perchloric acid digestion method of multielement tissue analysis. *Commun. Soil Sci. Plant Anal.* 3:425–436.

MICROWAVE DIGESTION OF PLANT TISSUE IN AN OPEN VESSEL

7

Yash P. Kalra and Douglas G. Maynard

INTRODUCTION

Electromagnetic radiation of frequencies of 100 to 100,000 megacycles per second is commonly referred to as microwave radiation. Samples are heated by the oscillating electromagnetic field. Radiation passes through glass or plastic and does not couple with the container material (as is the case with conventional heating). Because the radiation energy is applied directly to the digestion mixture, it provides extremely rapid heating and better control of power and time in the digestion of plant material by acid oxidation (White and Douthit, 1985).

EQUIPMENT

1. A commercially available laboratory microwave drying/digestion oven, such as Model MDS-81 D™ (CEM Corp., Indian Trail, NC) (Figure 7.1).
2. Teflon digestion vessels (with Teflon screw caps) of 120-mL capacity (CEM Corp., Indian Trail, NC).
3. Brinkmann dispensette acid dispensers, adjustable from 0 to 10 mL, for nitric acid (HNO$_3$) and hydrochloric acid (HCl).

1-57444-124-8/98/$0.00+$.50

63

FIGURE 7.1   Microwave Model MDS-81 D (CEM Corp., Indian Trail, NC).

4. Auto-pipette for hydrogen peroxide (H_2O_2).
5. Filter funnels.
6. Whatman No. 42 filter paper.

REAGENTS

1. Nitric acid [70% HNO_3 (specific gravity 1.42)], concentrated.
2. Hydrochloric acid [37% HCl (specific gravity 1.18)], concentrated.
3. Hydrogen peroxide (H_2O_2), 30%.

PROCEDURE

1. Transfer 0.30 to 0.40 g (0.01 g accuracy) plant tissue sample (20-mesh) into the microwave digestion vessel. Add 10 mL HNO_3 and swirl the vessel gently so that all the tissue comes in contact with the acid.
2. Screw on the cap. Do not use an insert in the cap. Load digestion vessels

on the turntable and put the turntable in the oven. Make sure that center wheel of turntable sits inside the tabs on the drive lugs. Switch on the turntable and check to ensure that assembly rotates smoothly.

3. Enter in time (30 minutes) and power (90%), press START, making sure that the exhaust is on FULL power and fumehood is on FAST function.
4. At the end of the digestion cycle, stop the turntable rotation. Leave the digestion vessels in the microwave oven for about 5 minutes to exhaust fumes.
5. Take digestion vessels out of microwave oven, carefully remove the cap under a fumehood, and slowly add 1.0 mL H_2O_2, and let stand for about 5 minutes.
6. Place the digestion vessels back into the microwave oven, start the turntable, and digest at 90% power for 15 minutes.
7. After cooling for about 5 minutes, remove the digestion vessel from the microwave oven, remove the cap under a fumehood, add 2.0 mL HCl, and let sit for about 5 minutes.
8. Place the digestion vessels back into the microwave oven, start the turntable, digest at 30% power for 10 minutes.
9. Remove the digestion vessels from the microwave oven, remove the cap (in a fumehood), and rinse with water. Rinse down sides of container.
10. Filter sample solutions (using Whatman No. 42 filter paper) into 100-mL volumetric flasks (in a fumehood).
11. Rinse digestion vessels three times to ensure that material is quantitatively transferred to funnels (make sure that it has filtered before second and third additions). Make up to 100 mL with deionized water.
12. After thorough mixing, transfer an aliquot into a 60-mL Nalgene bottle for determination of calcium (Ca), magnesium (Mg), potassium (K), sodium (Na), manganese (Mn), iron (Fe), aluminum (Al), phosphorus (P), and sulfur (S) by inductively coupled plasma atomic emission spectrometry (ICP-AES).

Note: The ICP-AES has its own computer. The weight and volume of each sample are entered and internal calibration and calculation are done with the blank subtracted (Hogan and Maynard, 1984).

COMMENTS

1. Immediately before use, all glassware, plasticware, and Teflon digestion vessels should be thoroughly rinsed, first with dilute HCl (1+3) and then with double-distilled water.

2. Screw caps are used to provide reflux action.
3. Reagents should be added to the samples in the fumehood.
4. The digestion vessel carousel should always be rotated during the digestion period, which ensures that all samples are subjected to the same microwave flux.
5. After HNO_3 digestion, samples must be cooled before adding H_2O_2; otherwise, there is excessive frothing due to the reaction between H_2O_2 and the hot acid digest.
6. It is essential that the filtrate does not have any particles that could clog the ICP-AES sample nebulizer.
7. Sodium in filter paper can impair delicate measurements unless removed before filtering the digests (Ali and Kalra, 1974).

TABLE 7.1   Results (mg/kg) obtained by microwave digestion method compared with National Institute of Standards and Technology values (Kalra et al., 1989)

Element	NIST citrus leaves SRM 1572		NIST pine needles SRM 1575	
	Microwave (n = 49)	NIST	Microwave (n = 42)	NIST
Calcium (Ca)	31,300 ± 1,030 (3.3)[a]	31,500 ± 1,000	4,160 ± 271 (6.5)	4,100 ± 200
Magnesium (Mg)	5,530 ± 150 (2.7)	5,800 ± 300	1,120 ± 47 (4.2)	1,200 ± 100
Potassium (K)	18,100 ± 660 (3.6)	18,200 ± 600	3,640 ± 93 (2.6)	3,700 ± 200
Sodium (Na)	154 ± 19 (12.3)	160 ± 20	16 ± 12 (75.0)	26 ± 9
Manganese (Mn)	20.7 ± 1.3 (6.3)	23 ± 2	667 ± 28 (4.2)	675 ± 15
Phosphorus (P)	1,340 ± 44 (3.3)	1,300 ± 200	1,190 ± 54 (4.5)	1,200 ± 200
Sulfur (S)	3,880 ± 106 (2.7)	4,070 ± 90	1,130 ± 39 (3.4)	1,180 ± 13
Iron (Fe)	75 ± 13 (17.3)	90 ± 10	140 ± 16 (11.4)	200 ± 10
Aluminum (Al)	76 ± 15 (19.7)	92 ± 15	401 ± 30 (7.5)	545 ± 30

[a] Coefficient of variation (%).

8. The microwave oven should be checked routinely for leakage using an electromagnetic monitor.
9. A microwave oven can digest 12 samples at a time, allowing digestion of 36 to 48 samples per day. For quality control, a minimum of one blank and one reference sample should be analyzed daily. Duplicates are done on approximately 5% of the samples.
10. Run a calibration curve.
11. As a measure of precision and accuracy, the mean concentration of elements (mg/kg) with standard deviation and coefficient of variation (%, in parentheses), as determined by the above method are compared with National Institute of Standards and Technology (NIST) values as given in Table 7.1. The Fe and Al results were the most problematic in precision and accuracy. Analysis of NIST standards (wheat flour, citrus and tomato leaves, and pine needles) by the proposed method gave lower results than the certified values; however, they compare well with the data obtained by other investigators (Kalra et al., 1989).

ACKNOWLEDGMENT

Reprinted with permission from the Canadian Forest Service (*Source:* Y.P. Kalra and D.G. Maynard. 1991. Methods Manual for Forest Soil and Plant Analysis. Inf. Rep. NOR-X-319. Forestry Canada, Northwest Region, Northern Forestry Centre, Edmonton, Alberta, Canada).

REFERENCES

Ali, M.W. and Y.P. Kalra. 1974. Sodium contamination by filter paper. *J. Assoc. Off. Anal. Chem.* 57:762.

Hogan, G.D. and D.G. Maynard. 1984. Sulphur analysis of environmental materials by vacuum inductively coupled plasma atomic emission spectrometry (ICP-AES), pp. 676–683. In: Proceedings Sulphur-84, International Conference, Calgary, Alberta, June 1984. Sulphur Development Institute of Canada, Calgary, Alberta.

Kalra, Y.P., D.G. Maynard, and F.G. Radford. 1989. Microwave digestion of tree foliage for multi-element analysis. *Can. J. For. Res.* 19:981–985.

White, Jr., R.T. and G.E. Douthit. 1985. Use of microwave oven and nitric acid-hydrogen peroxide digestion to prepare botanical materials for elemental analysis by inductively coupled argon plasma emission spectroscopy. *J. Assoc. Off. Anal. Chem.* 68:766–769.

MICROWAVE DIGESTION OF PLANT TISSUE IN A CLOSED VESSEL

8

Robert O. Miller

SCOPE AND APPLICATION

The method prepares plant tissue for the quantitative determination of the concentration of boron (B), calcium (Ca), copper (Cu), iron (Fe), magnesium (Mg), manganese (Mn), molybdenum (Mo), phosphorus (P), potassium (K), sodium (Na), sulfur (S), and zinc (Zn), using a nitric acid/hydrogen peroxide (HNO_3/H_2O_2) digestion mixture in conjunction with microwave heating in closed Teflon vessels (White and Douthit, 1985; Anderson and Henderson, 1986; Kalra et al. 1989; Stripp and Bogen, 1989). Digest analyte concentrations can be determined by atomic absorption spectrometry (AAS) (Watson and Isaac, 1990) and inductively coupled plasma atomic emission spectrometry (ICP-AES) (Keller, 1992).

The digestion procedure is based on the method described by Kingston and Jassie (1986, 1988), using HNO_3 and modified by Sah and Miller (1992), using concentrated HNO_3 and 30% H_2O_2. Digestion is facilitated by the application of microwave power and elemental volatilization is avoided using closed digestion vessels. Concentrations of these elements are used for plant nutrition diagnostics. However, the digest method is incomplete relative to the total oxidation of organic carbon (C). Calcium, Cu, Fe, K, Mg, Mn, Na, and Zn can be analyzed by AAS or ICP-AES. Boron, P, and S analyses require an ICP-AES with a vacuum spectrometer. Microwave HNO_3/H_2O_2 digests may not provide

100% recovery of aluminum (Al), silicon (Si), and selenium (Se). The method has a detection limit of approximately 0.01% for Ca, K, Mg, and P and 0.2 mg/ g for B, Cu, Fe, Mn, Mo, and Zn (sample dry basis). The method can also be used for the determination of trace elements [cadmium (Cd), lead (Pb), nickel (Ni), etc.] and is generally reproducible within ±7.0% for all analytes.

EQUIPMENT

1. Analytical balance, 250-g capacity, resolution ±0.1 mg.
2. Microwave digestion system and Teflon double wall digestion vessels (equipped with 200 psi relief seals).
3. Repipette dispensers, calibrated to 0.5 ± 0.05 mL and 2.0 ± 0.08 mL.
4. Polypropylene centrifuge tube with cap, 15 mL, graduated.
5. AAS and/or ICP-AES (vacuum or purged system).

REAGENTS

1. Deionized water, ASTM Type 1 grade.
2. Micro® Clean detergent.
3. Nitric acid (12N HNO$_3$), trace metal grade.
4. Hydrogen peroxide (H$_2$O$_2$), 30% solution.
5. Standard solutions (B, Ca, Cu, Fe, K, Mg, Mn, Mo, Na, P, S, and Zn): 1,000 mg/L obtained from commercial sources.
6. Multielement calibration standards (prepared from 1,000 mg/L standard solutions):
   • Ca, Mg, and K ranging from 5 to 500 mg/L
   • Na, P, and S ranging from 1.0 to 100 mg/L
   • B, Cu, Fe Mn, Mo, and Zn ranging from 0.10 to 10.0 mg/L
   Dilute solutions with 5% HNO$_3$.

PROCEDURE

1. Weigh 250 ± 5.0 mg dry plant tissue (see Comments 1, 2, and 3) and place in 120-mL Teflon digestion vessel. Include a method blank. For samples requiring Cu and Mo analyses, sample size should be increased to 500 ± 5.0 mg (dilution factor 30:1).
2. Using repipettes, add 0.5 mL trace metal grade concentrated HNO$_3$ and 2.00 mL 30% H$_2$O$_2$ to each vessel (see Comments 4 and 5). Ensure that the sample is completely wetted by the reagents.

3. Place digestion vessel in outer body shell, cap, and allow the sample and reagents to predigest for 30 minutes.
4. Close vessel (see Comments 6 and 7) relief valve and place the vessel (the microwave oven will accommodate 12 vessels) in the microwave oven and set the program for 4 minutes of 296 watts power and 8 minutes of 565 watts power.
5. At completion, remove the vessel from the microwave oven and place in a fumehood to cool (optional: place in freezer to cool for 30 minutes). In the fumehood, vent the vessel by rotating the release valve one half revolution. Vent until the vessel is completely depressurized. Remove the cap and rinse the cap into the vessel with deionized water.
6. Quantitatively transfer the contents of the digestion vessel into a centrifuge tube, dilute to 15-mL volume, cap the centrifuge tube, invert three times, and store (see Comments 8, 9, and 10).
7. Elemental analysis of the plant digest can be made using AES, AAS, ICP-AES, or other analytical methods. The method chosen will determine specific matrix modifications, calibration standards used, and the need for instrument-specific sample preparations and dilutions. Adjust and operate instruments in accordance with manufacturer's instructions. Calibrate instrument using calibration solutions. Determine the analyte concentrations of a method blank, unknown samples, and record concentrations in mg/L.

CALCULATIONS

Report Ca, K, Mg, Na, P, and S results to the nearest 0.001%:

$$\% \text{ analyte} = \frac{(\text{mg/L} - \text{method blank}) \times (60) \times (0.0001)}{\text{dry matter (\%)}/100}$$

Report B, Fe, Mn, and Zn results to the nearest 1 mg/kg;,Cu and Mo to the nearest 0.1 mg/kg:

$$\text{mg/kg analyte} = \frac{(\text{mg/L} - \text{method blank}) (60)}{\text{dry matter (\%)}/100}$$

COMMENTS

1. Teflon PFA 120-mL digestion vessel liners should be cleaned according to the following procedure: (1) soak liners in 1% solution of Micro®

Clean detergent for 1 hour, (2) rinse vessels in tap water, (3) rinse in solution of 0.5N HCl, (4) three deionized water rinses (ASTM Type 1 grade), and (5) dry for 1 hour at 80°C.

2. Plant tissue must be ground to pass a 40-mesh screen (<400 mM opening) to ensure homogeneity. Sample sizes less than 200 mg may lack homogeneity.

3. Plant tissues high in starch (i.e., cereal flours) may react violently and result in rupture seal failure. When digesting these materials, reduce sample mass to 200 mg of sample material. Examine digest for undecomposed plant tissue. Redigest sample if (1) significant residual particulate is noted in the digest or (2) the sample shows significant discoloration (i.e., gray or black, etc.).

4. Check repipette dispensing volume and calibrate using an analytical balance.

5. When adding reagent to vessels always wear protective clothing (i.e., eye protection, lab coat, disposable gloves and shoes). Always handle reagents and opening of vessels in an acid fumehood capable of high air flow at 100 cfm.

6. Inspect vessel rupture seal in the cap for replacement.

7. Follow microwave manufacturer's instructions for microwave lower calibration. Applying excessive microwave power may result in rupture of vessel seal or vessel wall failure.

8. Centrifuging may be necessary to clear the digest.

9. Samples having analyte concentrations exceeding the highest standard will require dilution and reanalysis.

10. Place 3.0 mL concentrate Micro® Clean Detergent in digestion vessel and allow to stand 30 minutes. Rinse out any particulate and finish cleaning according to set vessel cleaning procedure.

11. The technique is for the determination of the elements P, K, Ca, Mg, S, Na, B, Cu, Fe, Mn, and Zn.

REFERENCES

Anderson, D.L. and L.J. Henderson. 1986. Sealed chamber digestion for plant nutrient analysis. *Agron J.* 78:937–939.

Kalra, Y.P., D. Maynard, and F.G. Radford. 1989. Microwave digestion of tree foliage for multielement analysis. *Can. J. For. Res.* 19:981–985.

Keller, P.N. 1992. An overview of analysis by inductively coupled plasma atomic emission spectrometry, pp. 601–629. In: Akbar Montaser and D.W. Golightly (Eds.), *Inductively Coupled Plasma-Atomic Emission Spectrometry.* VCH Publishers, New York.

Kingston, H.M. and L.B. Jassie. 1986. Microwave energy for acid decomposition at elevated temperatures and pressures using biological and botanical samples. *Anal. Chem.* 58:2534–2541.

Kingston, H.M. and L.B. Jassie. 1988. Monitoring and predicting parameters in microwave dissolution, pp. 97–148. In: H.M. Kingston and L.B. Jassie (Eds.), *Introduction to Microwave Sample Preparation.* American Chemical Society, Washington, D.C.

Sah, R.N. and R.O. Miller. 1992. Spontaneous reaction for acid dissolution of biological tissue in closed vessels. *Anal. Chem.* 64:230–233.

Stripp, R.A. and D. Bogen. 1989. The rapid decomposition of biological materials by using microwave acid digestion bomb. *J. Anal. Toxic.* 13:57–59.

Watson, M.E. and R.A. Isaac. 1990. Analytical instruments for soil and plant analysis, pp. 691–740. In: R.L. Westerman (Ed.), *Soil Testing and Plant Analysis.* Third edition. SSSA Book Series 3. Soil Science Society of America, Madison, WI.

White, Jr., R.T. and G.E. Douthit. 1985. Use of microwave oven and nitric acid-hydrogen peroxide digestion to prepare botanical materials for elemental analysis by inductively coupled argon plasma emission spectroscopy. *J. Assoc. Off. Anal. Chem.* 68:766–769.

DETERMINATION OF TOTAL NITROGEN IN PLANT TISSUE

9

Donald A. Horneck and Robert O. Miller

INTRODUCTION

The determination of nitrogen (N) in organic materials is a procedure that has been around since the 1800s, when Johan Kjeldahl published his article in *Analytical Chemistry* (Morries, 1983; Scarf, 1988) and Dumas published his article in 1831. Kjeldahl developed a wet oxidation digestion that until recently had been the industry standard (Bradstreet, 1965; Nelson and Sommers, 1980; Isaac and Johnson, 1976; Jones, 1991; Wright and Wilkinson, 1993). Dumas devised a method for total N, using combustion. The advent of automated N analyzers, which utilize combustion for N determinations, in the last 10 years has begun to replace the Kjeldahl digestion method (McGeehan and Naylor, 1988; Hansen, 1989; Schmitter and Rihns, 1989). Nitrogen analyzers using combustion are typically less labor- and chemically intensive than the Kjeldahl method.

The Kjeldahl method uses sulfuric acid (H_2SO_4), a variety of catalysts, and salts to convert organically bound N in plant tissue to ammonium (NH_4) with its subsequent measurement. The Kjeldahl procedure is the official method of the American Association of Official Cereal Chemists (Anonymous, 1987) and the Association of Official Analytical Chemists (Helrich, 1995). The Kjeldahl procedure has several variances, mainly micro and macro, based primarily on sample size and required apparatus. The macro-Kjeldahl procedure was the

1-57444-124-8/98/$0.00+$.50
© 1998 by CRC Press LLC

original and required a large apparatus, frequently occupying an entire laboratory room. The macro-Kjeldahl uses a 1- to 2-g sample, 30 to 50 mL of acid and approximately 100 mL sodium hydroxide (NaOH) solution. The micro-Kjeldahl procedure, on the other hand, uses the same principles as the macro-Kjeldahl procedure, but the apparatus used is scaled down using less than a gram sample, 10 to 20 mL of acid, with the digestion typically carried out in a 40-hole aluminum block digester under a fumehood.

Catalysts that can be used are mercury (Hg), copper (Cu), selenium (Se), chromium (Cr), and titanium (Ti) (Simonne et al., 1993). The most common and probably the least toxic is the Ti and Cu mixture; the most toxic is Hg. Catalysts in various combinations can be purchased preground with potassium sulfate (K_2SO_4) as "Kel-tabs." Catalyst choice does not appear to make a large difference in N recovery for most routine analyses of plant tissues. However, the selection of the catalyst can affect length of time needed to complete the digestion (Nelson and Sommers, 1983; Simonne et al., 1993).

The Kjeldahl procedure will not recover 100% of the N in most samples. Nitrate (NO_3) and nitrite (NO_2) are not recovered unless a predigestion procedure is conducted, which reduces NO_3 to NH_4. Some cyclic nitrogenous compounds such as nicotinic acid also will not be recovered because of their resistance to this digestion. Dry combustion N analyzers [LECO, Heraeus, Perkin-Elmer, and Carlo Erba Nitrogen Analyzers (Determinators)] are the most common, and will recover slightly to significantly more N than a Kjeldahl because all the N in a sample is recovered (Simonne et al., 1994). For plant tissue samples like potato petioles where NO_3 content may exceed 3%, there will be a significant difference in measured N content between Kjeldahl N without predigestion and combustion-determined N.

The analysis of NH_4 after a Kjeldahl digestion can be performed by ammonium electrode, a continuous flow autoanalyzer, or by steam distillation. The choice for NH_4 determination will depend on sample volume, available facilities, and economics.

KJELDAHL NITROGEN: MICRO-KJELDAHL

Application

The Kjeldahl method quantitatively determines NH_4 and protein N in plant tissues based on the wet oxidation of organic matter using H_2SO_4 and a digestion catalyst (Isaac and Johnson, 1976). Ammonium may be determined by distillation into boric acid (H_3BO_3) and titration (Bremner and Keeney, 1965; Jones, 1991), spectrophotometric measurement (Smith, 1980; Baethegen and

Alley, 1989), specific ion electrode (Bremner and Tabatabai, 1972; Eastin, 1976; Powers et al., 1981), or diffusion conductivity (Carlson, 1978). The method does not quantitatively recover N from heterocyclic rings (such as nicotinic acid) or from oxidized forms such as NO_3 and NO_2. The method is used to assess plant N-sufficiency levels to estimate fertilizer needs (Chapman and Pratt, 1961). Detection limit for micro-Kjeldahl is approximately 0.05% N (dry sample basis) and is generally reproducible within ±4%.

Equipment

1. Analytical balance, 250-g capacity, resolution ±0.1 mg.
2. Acid fumehood.
3. Digestion heating block (400°C).
4. Volumetric digestion tubes, 75 mL.
5. Repipette dispenser, calibrated 3.0 ± 0.1 mL.

Reagents

1. Deionized water, ASTM Type 1 grade.
2. **Digest Catalyst Accelerator:** prepared by mixing (100:10:1) 100 g potassium sulfate (K_2SO_4), 10 g anhydrous copper sulfate ($CuSO_4$), and 1 g Se metal powder. This can be purchased as a prepared material under the brand name "Kjel-tab", "*Kjeltab*" or "*Kelmate*" distributed by various chemical suppliers.
3. Concentrated sulfuric acid (H_2SO_4), reagent grade.
4. Hydrogen peroxide (25 to 30% H_2O_2): use fresh because this material rapidly decomposes.
5. **Standard Calibration Solutions of Ammonium-Nitrogen (NH_4-N):** prepare six calibration standards ranging from 0.2 to 40.0 mg NH_4-N/L concentration diluted with 4% (v/v) H_2SO_4 prepared from 1,000 mg NH_4-N/L standard solution [weigh 4.7168 g dry ammonium sulfate $(NH_4)_2SO_4$ into a 1-L volumetric flask and dilute to volume with deionized water].

Procedure

1. Determine the moisture content of the plant tissue on a subsample.
2. Weigh 250.0 ± 15.0 mg air dried plant tissue (see Comment 1) and place in a 75-mL volumetric digestion tube (50-mL or 100-mL digestion tubes may be substituted). Include a method blank.
3. Add "Kjel-tab" or a 2-g scoop of catalyst and 6.0 mL concentrated H_2SO_4 (see Comments 2 and 3).

4. Mix on a vortex stirrer 15 seconds to thoroughly wet the sample with acid. *Note:* it is essential that all dry sample material be completely moistened by acid and well mixed to ensure complete digestion.

5. Place the digestion tube on a digestion block, preheated to 370°C for 30 seconds or long enough to achieve complete plant tissue break-up.

6. Remove the tube from the digestion block and carefully (slowly) add 2 to 5 mL 30% H_2O_2 in 1-mL increments to each digestion tube until the digest begins to clear. Because this reaction takes place very rapidly, slow additions should avoid excessive foaming. The addition of H_2O_2 can be excluded if very gradual heat increases are used until the sample stops foaming. Once foaming ceases, the digestion temperature can be brought to 370°C. Excluding this step may result in tubes foaming over and the need to reweigh and restart the digestion.

7. Place the digestion tube back on the digestion block maintained at 370°C for 2 hours. If excessive foaming occurs, remove from heat, cool 2 minutes, and add an additional 1 to 2 mL H_2O_2. At completion, a blue-green color may persist.

8. Remove the digestion tube from block and leave under the fumehood for 5 to 10 minutes. Then add 10 to 20 mL deionized water to each tube using a wash bottle to prevent hardening and crystal formation. Dilute digestion tubes to volume with deionized water, cap, and invert three times.

9. Sample digests can be analyzed for NH_4-N by three standard methods, conventional NH_4 distillation into H_3BO_3 and titration, which is described here (Jones, 1991), spectrophotometric determination of NH_4 [manual (Baethegen and Alley, 1989) or automated (Isaac and Johnson, 1976)], or diffusion-conductivity method of Carlson and coworkers (1978, 1990). Determine NH_4 concentration of a method blank, unknown samples and record results as mg NH_4-N/L in the digest (see Comments 4 and 5).

MANUAL DISTILLATION AND TITRATION

Reagents

1. **Mixed Indicator:** dissolve 0.3 g bromocresol green and 0.165 g methyl red indicators in 400 mL 95% ethanol, and bring to 500-mL volume.

2. **Boric Acid Indicator (4% H_3BO_3):** dissolve 20 g reagent grade H3BO3 in about 900 mL distilled water, heat, and swirl until dissolved. Add 20 mL mixed indicator (Reagent 1). Adjust to reddish-purple color with NaOH or HCl. This point is indicated when 1 mL tap water turns 1 mL indicator

solution a light-green, pH around 5.0, and dilute to 1 L with deionized water.

3. **Sodium Hydroxide (40% NaOH):** dissolve 400 g NaOH pellets in about 500 mL distilled water. Cool and bring to 1 L volume.

4. **Hydrochloric Acid (0.1N HCl):** pipette 8.3 mL concentrated HCl into 500 mL distilled water, and then bring to 1 L volume.

5. **Hydrochloric Acid (0.01N HCl):** dilute 100 mL 0.1N HCl with distilled water to a volume of 1 L.

Procedure

1. Turn on the heating unit to boiling flask and the flow of water through the condensers.

2. Pipette 10 mL Boric Acid Indicator Solution into a 125-mL Erlenmeyer flask. Place the Erlenmeyer flask under the condenser tip of the Kjeldahl unit. The end of the condenser should be in the boric acid indicator. Make sure the system is boiling before attaching the Kjeldahl flask to the distillation system in Step 3.

3. Quantitatively transfer the contents of the 75-mL volumetric digestion tube into a 300-mL Kjeldahl flask and attach to distillation system.

4. Add 30 mL 40% NaOH to the digested solution through the stopcock, rinse with a small amount of distilled water, and close the stopcock.

5. Distill approximately 75 mL into the 125-mL Erlenmeyer flask containing the Boric Acid Indicator Solution. Remove the steam bypass plug and then remove the Erlenmeyer flask.

6. Titrate with 0.1N HCl to a pink endpoint.

7. Make a blank determination on sample that was digested with each set of samples following the same procedure, without adding plant material.

Comment: Some of the reagents used in the Kjeldahl distillation determinations have been modified from the method presented by Bremner and Mulvaney (1982). These modifications have been developed so that the procedure can be used for routine plant analysis.

Calculations

1. Report total Kjeldahl nitrogen results to the nearest 0.01%:

$$\% \text{ N} = \frac{\text{mg/L NH}_4\text{-N in digest} - \text{method blank}) \times (0.075) \times (100)}{(\text{sample size, mg}) \times [\text{dry matter content } (\%)]/100}$$

Comments

1. Use 500 mg of sample if N content is less than 0.500%.
2. Check repipette dispenser delivery volume, recalibrate using an analytical balance.
3. When adding reagent to vessels and handling digests, always wear protective clothing (i.e., eye protection, lab coat, and disposable gloves and shoes). Always handle reagents and opening of vessels in an acid fumehood capable of high air flow at 100 cfm.
4. Samples having NH_4-N concentrations exceeding the highest standard will require dilution and reanalysis where colorimetric and conductive methods are used.
5. Sulfuric acid digest is classified as a hazardous waste and must be disposed of in a suitable manner.

References

Anonymous. 1987. *Approved Methods of the AACC.* American Association of Cereal Chemists, St. Paul, MN.

Baethegen, W.E. and M.M. Alley. 1989. A manual colorimetric procedure for measuring ammonium nitrogen in soil and plant Kjeldahl digests. *Commun. Soil Sci. Plant Anal.* 20:961–969.

Bradstreet, R.B. 1965. *The Kjeldahl Method for Organic Nitrogen.* Academic Press, New York.

Bremner, J.M. and D.R. Keeney. 1965. Steam distillation methods for the determination of ammonium, nitrate, and nitrite. *Anal. Chem.* 32:485–495.

Bremner, J.M. and M.A. Tabatabai. 1972. Use of an ammonia electrode for determination of ammonium in Kjeldahl analysis of soils. *Commun. Soil Sci. Plant Anal.* 3:159–165.

Bremner, J.M. and C.S. Mulvaney. 1982. Total nitrogen, pp. 595–624. In: A.L. Page, R.H. Miller, and D.R. Kenney (Eds.), *Method of Soil Analysis.* Part 2. Agronomy Monograph No. 9. American Society of Agronomy, Madison, WI.

Carlson, R.M. 1978. Automated separation and conductiometric determination of ammonia and dissolved carbon dioxide. *Anal. Chem.* 48:1528–1531.

Carlson, R.M., R.I. Cabrera, J.L. Paul, J. Quick, and R.Y. Evans. 1990. Rapid direct measurement of ammonium and nitrate in soil and plant tissue extracts. *Commun. Soil Sci. Plant Anal.* 21:1519–1529.

Chapman, H.D. and P.F. Pratt. 1961. *Methods of Analysis for Soils, Plants, and Waters.* Priced Publication 4034. Division of Agriculture Sciences, University of California, Berkeley.

Eastin, E.F. 1976. Use of ammonia electrode for total nitrogen determination in plants. *Commun. Soil Sci. Plant Anal.* 7:477–481.

Hansen, B. 1989. Determination of nitrogen as elementary N, an alternative to Kjeldahl. *Acta Agric. Scand.* 39:113–118.

Helrich, K. (Ed.). 1995. *Official Methods of Analysis of the AOAC*. Association of Official Analytical Chemists, Arlington, VA.

Isaac, R.A. and W.C. Johnson. 1976. Determination of total nitrogen in plant tissue. *J. Assoc. Off. Anal. Chem.* 59:98–100.

Jones, Jr., J.B. 1991. *Kjeldahl Method for Nitrogen (N) Determination*. Micro-Macro Publishing, Athens, GA.

McGeehan, S.L. and D.V. Naylor. 1988. Automated instrumental analysis of carbon and nitrogen in plant and soil samples. *Commun. Soil Sci. Plant Anal.* 19:493–505.

Morries, P. 1983. A century of Kjeldahl (1883–1983). *J. Assoc. Public Anal.* 21:53–58.

Nelson, D.W. and L.L. Sommers. 1980. Total nitrogen analysis of soil and plant tissues. *J. Assoc. Off. Anal. Chem.* 63:770–778.

Powers, R.F., D.L. van Gent, and R.F. Townsend. 1981. Ammonia electrode analysis of nitrogen in micro-Kjeldahl digests of forest vegetation. *Commun. Soil Sci. Plant Anal.* 12:19–30.

Scarf, H. 1988. One hundred years of the Kjeldahl method for nitrogen determination. *Arch. Acker. Pflanzenbau Bodenk.* 32:321–332.

Schmitter, B.M. and T. Rihns. 1989. Evaluation of a macrocombustion method for total nitrogen determination in feedstuffs. *J. Agric. Food Chem.* 37:992–994.

Simonne, E.H., J.B. Jones, Jr., H.A. Mills, D.A. Smittle, and C.G. Hussey. 1993. Influence of catalyst, sample weight, and digestion conditions on Kjeldahl N. *Commun. Soil Sci. Plant Anal.* 24:1609–1616.

Simonne, E.H., H.A. Mills, J.B. Jones, Jr., D.A. Smittle, and C.G. Hussey. 1994. Comparison of analytical methods for nitrogen analysis in plant tissues. *Commun. Soil Sci. Plant Anal.* 25:943–954.

Smith, V.R. 1980. A phenol-hypochlorite manual determination of ammonium-nitrogen in Kjeldahl digests of plant tissue. *Commun. Soil Sci. Plant Anal.* 11:709–722.

Wright, W.D. and D.R. Wilkinson. 1993. Automated Micro-Kjeldahl Nitrogen Determination: A Method. *Am. Environ. Lab.* 2/93:30–33.

AUTOMATED COMBUSTION METHOD

Scope and Application

This method quantitatively determines the amount of N in all forms (NH_4, NO_3, protein, and heterocyclic N) in plant tissues using an induction furnace and a thermal conductivity detector (Ebeling, 1968; McGeehan and Naylor, 1988; Hansen, 1989; Schmitter and Rihns, 1989). It is based on the method originally described by Dumas and later modified by Sweeney (1989), whereby samples are ignited in an induction furnace at approximately 900°C in helium (He) and oxygen (O_2) environment in a quartz combustion tube. An aliquot of the combustion gas is passed through a Cu catalyst to remove O_2 and convert nitrous oxides to N_2, scrubbed of moisture and carbon dioxide (CO_2), and N content determined by thermal conductivity. Similar instruments have the capability of the simultaneous analysis of carbon (C), hydrogen (H), or sulfur (S). The method is used to assess plant N-sufficiency levels. The method has a detection limit of 0.10% N (dry sample basis) and is generally reproducible to within ±5%.

Equipment

1. Analytical balance, 250-g capacity, resolution ±0.1 mg.
2. Total Nitrogen Analyzer: LECO, Carlo Erba, or Perkin-Elmer with induction furnace with thermal conductivity detector and operating supplies.
3. Tin foil encapsulating cups, which will depend on instrument available.
4. Desiccator, containing a desiccating agent.

Reagents

1. **Nitrogen Calibration Standards**
   - EDTA: 9.59% N
   - Glycine *p*-toluene sulfonate ($C_9H_{13}O_6SN$), 5.66% N
   - LECO Calibration Standard (PN 502-055), 2.40 ± 0.03% N

Procedure

1. Weigh 120 ± 5.0 mg dried plant material (see Comments 1 and 2) and place in a tared tin foil container, encapsulate, and record sample weight to the nearest 0.1 mg (see Comment 3).
2. Initialize the instrument following manufacturer's suggested protocol. Conduct a system leak check on combustion system. Perform blank stabilization test, analyze consecutive blanks until the blanks stabilize at a constant value.
3. Adjust and operate the instrument according to manufacturer's instructions, using calibration standards (provided by manufacturer or obtained commercially). Enter sample dry matter content and analyze unknown sample for total N. Report results to the nearest 0.01% nitrogen (see Comments 4, 5, and 6).

Calculation

Report sample N concentration to the nearest 0.01%.

Comments

1. Samples limited in material, should be dried over phosphorus pentoxide or magnesium perchlorate for 72 hours and analyzed without correction for moisture content or reported on as-received basis.
2. Specific instruments will have smaller or larger nominal sample size, check owner's manual for specifics.
3. Sample particulate must be ground to pass a 40-mesh screen (<400 μM) in order to assure adequate sample homogeneity.
4. Sample weight may be entered into instrument software using a balance interface.
5. Nitrogen content as determined by automated combustion method is generally slightly greater than values determined by the Kjeldahl method due to complete recovery of oxidized forms of N, such as NO_3 and NO_2, in addition to heterocyclic rings.
6. Standard addition techniques using N calibration standards can be used for the verification of the N content of unknown samples.

References

Ebeling, M.E. 1968. The Dumas method for nitrogen in feeds. *J. Assoc. Off. Anal. Chem.* 51:766–770.

Hansen, B. 1989. Determination of nitrogen as elementary N: An alternative to Kjeldahl. *Acta Agric. Scand.* 39:113–118.

McGeehan, S.L. and D.V. Naylor. 1988. Automated instrumental analysis of carbon and nitrogen in plant and soil samples. *Commun. Soil Sci. Plant Anal.* 19:493–505.

Schmitter, B.M. and T. Rihns. 1989. Evaluation of a macrocombustion method for total nitrogen determination in feedstuffs. *J. Agric. Food Chem.* 37:992–994.

Sweeney, R.A. 1989. Generic combustion method for determination of crude protein in feeds: Collaborative study. *J. Assoc. Off. Anal. Chem.* 72:770–774.

EXTRACTABLE NITRATE IN PLANT TISSUE: ION-SELECTIVE ELECTRODE METHOD

10

Robert O. Miller

SCOPE AND APPLICATION

The method semiquantifies the concentration of nitrate-nitrogen (NO_3-N) in plant tissues by extraction with an aluminum sulfate [$Al_2(SO_4)_3$] solution and subsequent determination by ion-selective electrode (ISE) (Baker and Smith, 1969; Milham et al., 1970; Carlson and Keeney, 1971; Mills, 1980; Baker and Thompson, 1992). The ISE determines NO_3-N by measuring an electrical potential developed across a thin layer of water-immiscible liquid or gel ion exchanger that is selective for NO_3 ions. This layer of ion exchanger is held in place by a porous membrane. The ISE is susceptible to interferences of chloride (Cl), bicarbonate (HCO_3), and sulfate (SO_4) anions and is sensitive to changes in solution ionic strength (i.e., high salt). Problems with precision have been noted by Mack and Sanderson (1971). The method has been used primarily to determine NO_3-N for assessing plant N fertility (Johnson and Ulrich, 1959; Chapman and Pratt, 1961; Maples et al., 1990). Generally, the method detection limit is approximately 200 mg/kg (sample dry basis) and is generally reproducible to within ±18.0%.

1-57444-124-8/98/$0.00+$.50

EQUIPMENT

1. Analytical balance, 250-g capacity, resolution ±0.001 g.
2. Repipette dispenser, calibrated to 25.0 ± 0.2 mL.
3. Reciprocating mechanical shaker, capable of 180 oscillations per minute.
4. Whatman No. 2V 11-cm filter paper or equivalent highly retentive paper.
5. Nitrate (NO_3) ISE.
6. pH/ion meter or pH-millivolt meter.

REAGENTS

1. Deionized water, ASTM Type 1 grade.
2. **Extracting/Ionic Strength Adjusting Solution** [0.01M $Al_2(SO_4)_3$, 0.02M H_3BO_3, 0.01M Ag_2SO_4, and 0.02M NH_2HCO_3]: dissolve 67 g aluminum sulfate [$Al_2(SO_4)_3$], 12 g boric acid (H_3BO_3), 20 g silver sulfate (Ag_2SO_4) and 19 g sulfamic acid (NH_2HCO_3) in water and dilute to 10 liters.
3. **Standard Nitrate Solutions:** to a 1,000-mL volumetric flask, add 0.7221 g oven dry potassium nitrate (KNO_3) and make up to volume with Extracting Solution. This yields a solution containing 100 ppm of NO_3-N. Prepare calibration standards from extraction solution of 5.0, 10.0, 15.0, 20.0, 30.0, and 50.0 mg NO_3-N/L.

PROCEDURE

1. Weigh 500.0 ± 1.0 mg air-dried plant tissue (see Comments 1, 2, and 3) and place in 50-mL extraction vessel.
2. Add 25.0 ± 0.2 mL Extracting Solution and place on reciprocating mechanical shaker for 30 minutes. Include a method blank.
3. Filter extract; refilter if filtrate is cloudy and retain for analysis.
4. Calibrate ISE/millivolt meter using standard calibration solutions and operate instrument in accordance with manufacturer's instructions. Develop calibration curve for the ISE using standards. Determine nitrate concentration of plant sample and record results as mg NO_3-N/L in extraction solution (see Comments 4 and 5).

CALCULATIONS

Report mg of NO_3-N in sample to the nearest 10 mg/kg:

$$mg/kg = \frac{(NO_3\text{-}N \text{ in extract mg/L} - \text{method blank}) \times (50)}{\text{dry matter content (\%)/100}}$$

COMMENTS

1. Plant tissue must be ground to pass 40-mesh screen (<0.40 mm) in order to ensure adequate homogeneity.
2. Sample mass may be adjusted in accordance with expected analyte concentrations. For materials containing <500 mg NO_3-N/kg, increase sample size to 1,000 mg.
3. Check repipette dispensing volume, calibrate using an analytical balance.
4. Routinely check ISE calibration every third sample using a mid-range standard. In specific instances, the ISE may be susceptible to radio frequency energy from surrounding electronic equipment (Carlson, 1992).
5. Samples having NO_3 concentrations exceeding the highest standard will require dilution and reanalysis.

REFERENCES

Baker, A.S. and R. Smith. 1969. Extracting solution for potentiometric determination of nitrate in plant tissue. *J. Agric. Food Chem.* 17:1284–1287.

Baker, W.H. and T.L. Thompson. 1992. Determination of nitrate-nitrogen in plant samples by selective ion electrode, pp. 25–31. IN: C.O. Plank (Ed.), *Plant Analysis Reference Procedures for the Southern Region of the United States.* Southern Cooperative Series Bulletin 368. University of Georgia, Athens.

Carlson, R.M. and D.R. Keeney. 1971. Specific ion electrodes for soil, plant, and water analysis, pp. 39–63. In: L.M. Walsh (Ed.), *Instrumental Methods for Analysis of Soils and Plant Tissue.* Soil Science Society of America, Madison, WI.

Carlson, R.M. 1992. Personal communication.

Chapman, H.D. and P.F. Pratt. 1961. Methods of Analysis for Soils, Plants and Waters. Priced Publication 4034. Division of Agriculture Sciences, University of California, Berkeley.

Johnson, C.M. and A. Ulrich. 1959. Analytical Methods for Use in Plant Analysis. Bulletin 766. Agricultural Experiment Station, University of California, Berkeley, CA.

Mack, A.R. and R.B. Sanderson. 1971. Sensitivity of the nitrate-ion membrane electrode in various oil extracts. *Can. J. Soil Sci.* 51:95–104.

Maples, R.L., W.N. Miley, and T.C. Keisling. 1990. Nitrogen recommendations for cotton based on petiole analysis: Strengths and limitations, pp. 59–63. In: W.N. Miley and D.M. Oosterhuis (Eds.), *Nitrogen Nutrition of Cotton: Practical Issues.* American Society of Agronomy, Madison, WI.

Milham, P.J., A.S. Awad, R.E. Paul, and J.H. Bull. 1970. Analysis of plants, soils, and waters for nitrate by using an ion-selective electrode. *Analyst* 95:751–759.

Mills, H.A. 1980. Nitrogen specific ion electrode for soil, plant, and water analysis. *J. Assoc. Off. Anal. Chem.* 63:797–801.

DETERMINATION OF AMMONIUM-NITROGEN IN PLANT TISSUE

11

Liangxue Liu

INTRODUCTION

Ammonium (NH_4) and nitrate (NO_3) are the two inorganic nitrogen (N) forms available for plant uptake. Fertilizer N, applied to soil in the NH_4 form, is rapidly oxidized to NO_3 via the nitrification process, but the rate of its conversion is dependent on a number of factors, including temperature and population of soil organisms. Hence, plants may be exposed to varying proportions of both inorganic N forms. In order to assess the relative proportion of NH_4 absorption and accumulation of NH_4 in the plant, an analysis for NH_4 is required. Water-extractable NH_4 has been used to show NH_4 accumulation when plants are subjected to only NH_4-N nutrition (Liu and Shelp, 1992).

EQUIPMENT

1. 90-mL glass vial with snap cap.
2. Reciprocating or rotating shaker capable of 250 rpm.
3. Whatman No. 42 filter paper.

1-57444-124-8/98/$0.00+$.50
© 1998 by CRC Press LLC

REAGENTS

1. Deionized high-resistance water (18 MΩ).
2. **Alkaline Phenol:** add 83.0 g liquefied phenol to about 800 mL deionized water. While cooling under tap water or in an ice bath, slowly add with swirling 96.0 g sodium hydroxide [(NaOH) 50% w/w]. Cool to room temperature, dilute to 1 L with deionized water, and mix well. Store in an amber glass container. This material is corrosive and stable for 2 weeks.
3. **Sodium Hypochlorite Solution:** dilute 86 mL 50% sodium hypochlorite (NaOCl) solution to 100 mL with deionized water and mix thoroughly. Commercially available bleach is 5.25% active. If used, only 82 mL are needed. Prepare fresh weekly.
4. **Sodium Nitroprusside Solution:** dissolve 1.1 g sodium nitroprusside in about 600 mL deionized water. Dilute to 1 L with deionized water and mix thoroughly. Store in an amber glass container. This material is stable for 1 month.
5. **Disodium EDTA Solution:** dissolve 1.0 g 50% (w/w) sodium hydroxide (NaOH) and 41.0 g Na_2EDTA to about 800 mL deionized water. Dilute to 1 L with deionized water. Add 3 mL Brij-35 and mix well.
6. **Preparation of Standard Solutions:** dissolve 4.7168 g accurately weighed ammonium sulfate [$(NH_4)_2SO_4$] in about 800 mL deionized water. Dilute to 1,000 mL with deionized water and mix thoroughly. This produces a stock standard of 1,000 mg NH_4-N/L. Pipette 10 mL stock standard into a 100-mL volumetric flask, dilute to volume with deionized water and mix thoroughly. This produces a working standard of 100 mg NH_4-N/L. Pipette 1, 2, 4, 6, 8, and 10 mL of working standard into individual 100-mL volumetric flasks. Dilute to volume with deionized water and mix thoroughly. The above solutions contain 1, 2, 4, 6, 8, and 10 mg NH_4-N/L, respectively.

PROCEDURE

1. Grind dried plant material to a fine powder.
2. Weigh 250 mg of the fine powder plant material into a 90-mL glass vial.
3. Add 25 mL deionized water and cap the vial.
4. Place the vial on a reciprocating shaker and shake at 250 rpm for 30 minutes.
5. Remove from the shaker and allow to stand for 15 minutes.
6. Filter through Whatman No. 42 filter paper into a plastic vial for NH_4 analysis.

ANALYSIS

The NH_4 in the filtrate can be analyzed using the Berthelot method on the TRAACS 800™ AutoAnalyzer, a third generation autoanalyzer (Tel and Heseltine, 1990). The sample is mixed with EDTA solution to eliminate the interference and precipitation of calcium (Ca), magnesium (Mg), and iron (Fe). The NH_4 is then dialyzed into a stream of alkaline phenol (actually sodium phenate). A color formation reaction occurs when an NH_4 salt is added to alkaline phenol followed by the addition of sodium hypochlorite and sodium nitroprusside. The intensity of the color complex is directly proportional to the concentration of NH_4 in the solution. The resultant indigo-blue dye absorbs strongly between 630 to 720 nm and is generally measured at 660 nm using a 10-mm flow cell.

A manual method of NH_4 determination has been described by Beathgen and Alley (1989).

COMMENTS

1. Colored organic or colloidal materials are removed by dialysis.
2. The determination of NH_4 in water extracts is rapid and reliable. But water may extract only free NH_4 present in the plant tissue.

NOTE

This chapter is dedicated to the late Mr. Dirk Tel for his outstanding contribution to the development of many analytical methods.

REFERENCES

Beathgen, W.E. and M.M. Alley. 1989. A manual colorimetric procedure for measuring ammonium nitrogen in soil and plant Kjeldahl digests. *Commun. Soil Sci. Plant Anal.* 20:961–969.

Liu, L. and B.J. Shelp. 1992. Nitrogen partitioning in greenhouse-grown broccoli in response to varying NH_4^+:NO_3^- ratios. *Commun. Soil Sci. Plant Anal.* 24:45–60.

Tel, D.A. and C. Heseltine. 1990. The analyses of KCl soil extracts for nitrate, nitrite and ammonium using a TRAACS 800 Analyzer. *Commun. Soil Sci. Plant Anal.* 21:1681–1688.

TOTAL SULFUR DETERMINATION IN PLANT TISSUE

12

C. Grant Kowalenko and Cornelis (Con) J. Van Laerhoven

INTRODUCTION

Numerous methods have been used for the determination of total sulfur (S) in plant material, with different combinations of converting all of the S to one form before the final quantification (Beaton et al., 1968). In some cases, the conversion and quantification has been combined into one instrument (e.g., commercially available S analyzers based on dry ashing) or determined without physical conversion by direct atomic methods (e.g., X-ray or neutron activation) (Watson and Isaac, 1990). In the methods with separate conversion and quantification, the quantification has usually been based either on barium precipitation or as the sulfide after reduction by hydriodic acid reagent. More recently, quantification has been done by inductively coupled plasma atomic emission spectrometry (ICP-AES) and ion chromatography (Artiola and Ali, 1990; Peverill, 1993; Sterrett et al., 1987).

Selection of an acceptable total S method for plant material has been limited by the availability of certified reference materials (Koch, 1989; Soon et al., 1996; Topper and Kotuby-Amacher, 1990; Zhao et al., 1994). Interlaboratory comparisons have shown that variation in plant total S measurements have been higher than that for analysis of most other elements (Peverill, 1993; Sterrett et al., 1987). It is interesting to note that in an earlier interlaboratory comparison, S was not included even though 10 elements (macro and micro) were consid-

1-57444-124-8/98/$0.00+$.50
© 1998 by CRC Press LLC

ered (Watson, 1981). For a number of National Institute of Standards and Technology (NIST) [formerly National Bureau of Standards (NBS)] plant reference materials (Orchard Leaves 1571, Spinach 1573, Pine Needle 1575, Apple Leaves 1515, Peach Leaves 1547, Corn Stalk 8412), S was either not reported or only a non-certified value was given for information. Recently, certified biological reference materials (Wheat Flour 8436, 8437, and 8438, Wheat Gluten 8418, Bovine Muscle 8414, Whole Egg 8415, and Whole Milk 8435) by NIST have included certified values for S but not for Corn Bran (8433), Corn Starch (8432), and Microcrystalline Cellulose (8416) (Ihnat, 1994a–c; Ihnat and Wolynetz, 1994). The methods used for S determinations in that study included acid digestion followed by ICP-AES, neutron activation, commercial combustion/determination instruments and various combustion procedures followed by detection involving spectrophotometry, ion chromatography, or gravimetry, but details of the methods used were not specifically documented.

The selection of the reference method for this publication focused on theoretical effectiveness for determining S specifically (as compared to multiple element methods) and the adaptability of the method by many different laboratories. Methods that were totally dependent on specific instrumentation, such as neutron activation or commercial combustion/detection instruments, were avoided largely because of cost. Commercial combustion/detection instruments (e.g., LECO, Carlo Erba, Antek) based on different combustion and/or detection characteristics have been evaluated for their effectiveness in determining total S in plant materials on their own, but have not been compared with one another (David et al., 1989; Hern, 1984; Jackson et al., 1985; Kirsten, 1979; Kirsten and Nordenmark, 1987; Matrai, 1989). These instruments are constantly being refined and many factors such as oven temperature, and addition and type of combustion-enhancing catalyst, etc. can influence the results, which makes comparisons of published values by these instruments very difficult to evaluate.

The method selected for this publication involves two parts: (1) decomposition of organic components of the sample and conversion of all the inorganic S to the sulfate (SO_4) form and (2) quantification of the SO_4. The criteria for selection of the decomposition/conversion step were effectiveness and compatibility with various methods for quantifying the SO_4 produced. Numerous methods have been developed for the decomposition/conversion step, mainly dry ashing, including closed-vessel oxygen-enhanced combustion/absorption, and wet digestion (Beaton et al., 1968). Wet digestion has predominantly been done with acids, but because volatile losses of S can occur with heating in acidic conditions (Hafez et al., 1991; Randall and Spencer, 1980; Zhao et al., 1994), this procedure is not recommended on a theoretical basis even if precautions such as careful temperature control were proposed. Acid digestion may be acceptable for situations where extreme accuracy for S determination may be compromised for multiple element analyses purposes (Pritchard and Lee, 1984; Wolf, 1982). Dry ashing was selected over closed-flask oxygen-enhanced combustion/absorption because dry ashing was considered simple and fast for large numbers of samples (Tabatabai et al., 1988) and the two methods have been shown to yield similar results (Johnson and Nishita, 1952). Dry ashing should be done in the presence of an alkali to prevent volatile losses of S, and various types of alkali have been used (Beaton et al., 1968). Although magnesium nitrate [$Mg(NO_3)_2$] has been used frequently (Cunniff, 1995; Guthrie and Lowe, 1984; Johnson and Nishita, 1952; Wolf, 1982) as the alkali in digestions, a procedure involving sodium bicarbonate ($NaHCO_3$) and silver oxide (Ag_2O) as the alkali was selected based on recent work that showed its adaptability with different methods of SO_4 quantification (Artiola and Ali, 1990; Lea and Wells, 1980; Perrott et al., 1991; Tabatabai et al., 1988).

The focus of the proposed method is on the combustion of the sample rather than on the quantification of the inorganic S produced. The final choice for the quantification procedure is not strictly prescribed, with several options available according to the instrumentation (automated or manual) available to the analyst. It is assumed that the dry-ashing procedure is complete regarding organic matter destruction and conversion of all the inorganic S to the SO_4 form and that barium-based, hydriodic acid reduction, ion chromatography, and ICP-AES methods will yield the same results if precautions are taken to eliminate interferences, etc. It should be realized, however, that barium-based and ion chromatography methods are specific to inorganic SO_4, hydriodic acid includes SO_4 of both organic and inorganic forms, and ICP-AES includes all S forms. The final method proposed is based on the method reported by Tabatabai et al. (1988). The method was simplified by combining $NaHCO_3$ and Ag_2O additions as a single addition of oxidation aid mixture.

EQUIPMENT

1. Analytical balance with 0.0001 g readability.
2. Muffle furnace to heat crucibles to 550°C and hold that temperature for 3 hours.
3. Hot plate programmed at 200°C.
4. Porcelain crucibles, 30-mL capacity.
5. Equipment needed to quantify S in sample digests by barium-based, hydriodic acid reduction, ion chromatography, or ICP-AES method.

Reagents

1. **Oxidation-Aid Mixture:** mix 2 g ACS grade silver oxide (Ag_2O) with 50 g ACS grade sodium bicarbonate ($NaHCO_3$).
2. **Ashing Guard Reagent:** ACS grade sodium bicarbonate ($NaHCO_3$).
3. **Ash-Dissolution Reagent:** prepare 1M acetic acid solution by diluting 58 mL of concentrated trace metal grade acid to 1 L with water.
4. **Calibration Standards:** dissolve 5.436 g ACS grade potassium sulfate (K_2SO_4) in 1 L water to prepare a 1,000 ppm sulfate-sulfur (SO_4-S) stock standard and dilute 10 times to obtain an intermediate 100 ppm SO_4-S standard. Prepare working standards at concentrations appropriate for the quantification method used.
5. Solutions for selected S quantification methods.

PROCEDURE

1. Weigh and record actual weight of 0.100 ± 0.005 g oven-dry finely ground (<100-mesh) plant tissue and place in porcelain crucible. Add 0.520 ± 0.005 g of the oxidation-aid mixture and mix with the sample. Carefully tap the crucible on the bench top to settle the mixed contents as an even layer in the bottom of the crucible. Add 0.500 ± 0.005 g $NaHCO_3$ ashing guard layer to completely cover the sample layer.
2. Prepare reagent blanks and appropriate standard solutions for matrix-matched calibrations as above. Aliquots of standard solutions should be dried (<100°C) in the crucibles prior to application of oxidation-aid mixture and guard layers.
3. Place crucibles in the muffle furnace programmed to heat to 550°C, hold this temperature for 3 hours, and then cool to room temperature.
4. Carefully add 15 mL of the ash-dissolution reagent to each crucible. To speed up the neutralization, removal of CO_2, and dissolution of ash, heat the crucibles on a hot plate adjusted to 200°C for 20 minutes. After crucibles cool, quantitatively transfer contents into volumetric flasks using a funnel and a stream of water from a wash bottle, then make up to volume with water. The final volume should be selected to accommodate the sensitivity and range of the quantification method (e.g., 100 mL for ion chromatography).
5. Determine solution SO_4-S by the selected quantification method.

6. Calculate plant total S concentration, taking into consideration the weight of the sample, volume of the dissolved ash, reagent blanks, etc.

ANALYSIS

Ion Chromatography Quantification

Since the prepared digest solution includes precipitates and insoluble components, solutions must be filtered prior to injection into the chromatography system to prevent system blockages and column fouling. A 0.45-mm syringe filter is recommended. Suppressed ion chromatography methods for similar types of analyses (Busman et al., 1983; Hafez et al., 1991; Tabatabai et al., 1988) can be used. As a starting point for establishing suitable suppressed anion analysis chromatographic conditions, employ instrumentation manufacturer's standard conditions; for example, use of a Dionex AG4A (4 × 50 mm) precolumn and AS4A (4 × 250 mm) analytical column in conjunction with anion micromembrane suppressor (AMMS), 2 mL/min flow of 1.7 mM NaHCO$_3$/1.8 mM Na$_2$CO$_3$ eluant, 3 mL/min flow of 25 mM H$_2$SO$_4$ suppressor regenerant, direct 50 μL sample injection and conductivity measurement at 30 mS full scale (Dionex Corporation, 1987). Non-suppressed single-column ion chromatographic methods (Abraham and DeMan, 1987; Barak and Chen, 1987; Maynard et al., 1987) can also be used.

Inductively Coupled Plasma Atomic Emission Spectrometry Quantification

General procedures and principles for ICP-AES operation (Munter et al., 1984; Watson and Isaac, 1990) and methods specific to measurements of S under conditions similar to those described for this ashing method (Hogan and Maynard, 1984; Kalra et al., 1989; Perrott et al., 1991; Pritchard and Lee, 1984; Zhao et al., 1994) should be adapted for use.

Hydriodic Acid Reduction Quantification

The hydriodic acid method originally proposed by Johnson and Nishita (1952) can be used for manual analyses with readily available laboratory supplies and equipment. The modifications to this original procedure have been successful in reducing the time required for each analysis (Kowalenko, 1985; Kowalenko and Lowe, 1972). The hydriodic acid reduction method outlined in Chapter 13 of this volume on SO_4 determination in plant tissue should be followed with appropriate modifications.

Barium-Based Quantification

A wide variety of barium-based methods have been used for SO_4 quantification (Beaton et al., 1968), and these could be manual (Guthrie and Lowe, 1984; Jones, 1996) or automated (Lea and Wells, 1980).

COMMENTS

On a theoretical basis and from laboratory observations that loss of S can occur when acidic solutions are heated, it may be desirable to avoid heating the crucibles of ashed materials in order to dissolve ash and neutralize $NaHCO_3$. Although gently swirling of crucible contents several times over a 2-hour period appears to accomplish the neutralization and release of CO_2, crucible contents could be left to stand at room temperature overnight before transfer to volumetric flasks. Use of HCl rather than acetic acid also accomplishes neutralization quickly without the need to heat crucible contents. In this case, care must be taken to avoid losses due to spattering associated with rapid CO_2 release. It may be prudent to wash crucible contents into the flask with a small volume of water prior to use of the acid and allow most of the reaction to take place in the volumetric flask.

Quality assurance measures such as matrix matching of sample digests and calibration standards and tests involving pre-ash spiking, i.e., the method of standard additions, should be undertaken to evaluate the appropriateness of any quantification method that is used. For example, in suppressed ion chromatography, peak area response per unit SO_4-S is affected by final solution acidity (Hafez et al., 1991); therefore, acid neutralization of the $NaHCO_3$ after ashing must be done as consistently and comparably as possible.

When S quantification is carried out by ion chromatography, analysts are encouraged to include phosphate (PO_4) in instrument calibration standards to ensure that selected chromatographic conditions not only resolve the SO_4 peak from the large acetate peak, but also completely separate PO_4 from SO_4-S. As a result, analysis time by suppressed ion chromatography may be lengthened marginally, but reducing eluent carbonate (CO_3) concentration slightly will enhance PO_4 and SO_4 peak displacement beyond the large matrix acetate peak, improve PO_4 and SO_4 separation and measurement precision. In our laboratory, lowering eluant CO_3 from 1.8 mM to 1.44 mM using a Dionex A54A column set still permitted analysis times of 10 minutes per sample and provided better peak resolution. Although Tabatabai et al. (1988) indicated HCl could not be used to neutralize and dissolve the ignition residue due to problems resolving the large Cl peak from the SO_4 peak, the previously mentioned chromatographic conditions permitted S quantification when HCl was substituted for acetic acid. If other oxyanionic sample components (such as selenate, selenite, arsenate, etc.) are expected to be present at substantial concentrations, standards prepared to include these potential interferents should be run to ensure these do not co-elute with SO_4.

For accurate analyses of S on its own compared to multiple element analyses using ICP-AES, various factors for measurement of this element should be optimized. Attention should be given to instrument settings such as wavelength selection, measurement height above the flame, etc. and solution matrix and sample characteristics (Hogan and Maynard, 1984; Kalra et al., 1989; Kovács et al., 1996; Perrott et al., 1991; Pritchard and Lee, 1984; Zhao et al., 1994).

The hydriodic acid reduction method is quite sensitive and free from interferences. There is a need for uniform water contents for samples and calibration standards during the digestion/distillation (Kowalenko and Lowe, 1972).

Barium-based methods are subject to interferences, and various methods have been used to minimize this problem. It may be necessary to decrease the final volume of the dissolved ash material to accommodate the relatively low sensitivity of these methods; e.g., the final volume for the ashed material by method of Lea and Wells (1980) was 20 mL.

REFERENCES

Abraham, V. and J.M. DeMan. 1987. Determination of total sulfur in canola oil. *J. Am. Oil Chem. Soc.* 64:384–387.

Artiola, J.F. and A.S. Ali. 1990. Determination of total sulfur in soil and plant samples using sodium bicarbonate/silver oxide, dry ashing and ion chromatography. *Commun. Soil Sci. Plant Anal.* 21:941–949.

Barak, P. and Y. Chen. 1987. Three-minute analysis of chloride, nitrate, and sulfate by single column anion chromatography. *Soil Sci. Soc. Am. J.* 51:257–258.

Beaton, J.D., G.R. Burns, and J. Platou. 1968. *Determination of Sulphur in Soils and Plant Material,* pp. 17–19. Technical Bulletin 14. The Sulphur Institute, Washington, D.C.

Busman, L.M., R.P. Dick, and M.A. Tabatabai. 1983. Determination of total sulfur and chlorine in plant materials by ion chromatography. *Soil Sci. Soc. Am. J.* 47:1167–1170.

Cunniff, P. (Ed.). 1995. *Official Methods of Analysis of AOAC International*, p. 22. AOAC International, Arlington, VA.

David, M.B., M.J. Mitchell, and D. Aldcorn. 1989. Analysis of sulfur in soil, plant and sediment materials: Sample handling and use of an automated analyzer. *Soil Biol. Biochem.* 21:119–123.

Dionex Corporation. 1987. *Dionex Ion Chromatography Cookbook.* Literature Part Number 032861. Sunnyvale, CA.

Guthrie, T.F. and L.E. Lowe. 1984. A comparison of methods for total sulphur analysis of tree foliage. *Can. J. For. Res.* 14:470–473.

Hafez, A.A., S.S. Goyal, and D.W. Rains. 1991. Quantitative determination of total sulfur in plant tissues using acid digestion and ion chromatography. *Agron. J.* 83:148–153.

Hern, J.L. 1984. Determination of total sulfur in plant materials using an automated sulfur analyzer. *Commun. Soil Sci. Plant Anal.* 15:99–107.

Hogan, G.D. and D.G. Maynard. 1984. Sulphur analysis of environmental materials by vacuum inductively coupled plasma atomic emission spectrometry (ICP-AES), pp. 676–683. In: Proceedings of Sulphur-84. Sulphur Development Institute of Canada (SUDIC), Calgary, Alberta.

Ihnat, M. 1994a. Characterization (certification) of Bovine Muscle Powder (NIST RM 8414), Whole Egg Powder (NIST RM 8415) and Whole Milk Powder (NIST RM 8435) Reference Materials for essential and toxic major, minor and trace element constituents. *Fresenius Z. Anal. Chem.* 348:459–467.

Ihnat, M. 1994b. Characterization (certification) of three wheat flours and a wheat gluten reference material (NIST RM 8436, 8437, 8438, and 8418) for essential and toxic major, minor and trace element constituents. *Fresenius Z. Anal. Chem.* 348:468–473.

Ihnat, M. 1994c. Characterization (certification) of Corn Bran (NIST RM 8433), Corn Starch (NIST RM 8432), and Microcrystalline Cellulose (NIST RM 8416) Reference Materials for essential and toxic major, minor and trace elemental constituents. *Fresenius Z. Anal. Chem.* 348:474–478.

Ihnat, M. and M.S. Wolynetz. 1994. An interlaboratory characterization (certification) campaign to establish the elemental composition of a new series of agricultural/food reference materials. *Fresenius Z. Anal. Chem.* 348:452–458.

Jackson, L.L., E.E. Engleman, and J.L. Peard. 1985. Determination of total sulfur in lichens and plants by combustion-infrared analysis. *Environ. Sci. Tech.* 19:437–441.

Johnson, C.M. and H. Nishita. 1952. Microestimation of sulfur in plant materials, soils and irrigation waters. *Anal. Chem.* 24:736–742.

Jones, Jr., J.B. 1996. Determination of total sulphur in plant tissue using the HACH kit spectrophotometer technique. *Sulphur Agr.* 19:58–62.

Kalra, Y.P., C.G. Maynard, and F.G. Radford. 1989. Microwave digestion of tree foliage for multi-element analysis. *Can. J. For. Res.* 19:981–985.

Kirsten, W.J. 1979. Automated methods for the simultaneous determination of carbon, hydrogen, nitrogen, and sulfur, and for sulfur alone in organic and inorganic materials. *Anal. Chem.* 51:1173–1179.

Kirsten, W.J. and B.S. Nordenmark. 1987. Rapid, automated, high-precision method for micro, ultramicro, and trace determinations of sulfur. *Anal. Chim. Acta* 196:59–68.

Koch, W.F. 1989. Ion chromatography and the certification of standard reference materials. *J. Chromat. Sci.* 27:418–421.

Kovács, B., Z. Gyôri, J. Prokisch, J. Loch, and P. Dániel. 1996. A study of plant sample preparation and inductively coupled plasma emission spectrometry parameters. *Commun. Soil Sci. Plant Anal.* 27:1177–1198.

Kowalenko, C.G. 1985. A modified apparatus for quick and versatile sulphate sulphur analysis using hydriodic acid reduction. *Commun. Soil Sci. Plant Anal.* 16:289–300.

Kowalenko, C.G. and L.E. Lowe. 1972. Observations on the bismuth sulfide colorimetric procedure for sulfate analysis in soil. *Commun. Soil Sci. Plant Anal.* 3:79–86.

Lea, R. and C.C. Wells. 1980. Determination of extractable sulfate and total sulfur from plant and soil material by an autoanalyzer. *Commun. Soil Sci. Plant Anal.* 11: 507–516.

Matrai, P.A. 1989. Determination of sulfur in ocean particulates by combustion fluorescence. *Marine Chem.* 26:227–238.

Maynard, D.G., Y.P. Kalra, and F.G. Radford. 1987. Extraction and determination of sulfur in organic horizons of forest soils. *Soil Sci. Soc. Am. J.* 51:801–806.

Munter, R.C., T.L. Halverson, and R.D. Anderson. 1984. Quality assurance for plant tissue analysis by ICP-AES. *Commun. Soil Sci. Plant Anal.* 15:1285–1322.

Perrott, K.W., B.E. Kerr, M.J. Kear, and M.M. Sutton. 1991. Determination of total sulphur in soil using inductively coupled plasma-atomic emission spectrometry. *Commun. Soil Sci. Plant Anal.* 22:1477–1487.

Peverill, K.I. 1993. Soil testing and plant analysis in Australia. *Aust. J. Exp. Agric.* 33:963–971.

Pritchard, M.W. and J. Lee. 1984. Simultaneous determinations of boron, phosphorus and sulphur in some biological and soil materials by inductively-coupled plasma emission spectrometry. *Anal. Chim. Acta* 157:313–326.

Randall, P.J. and K. Spencer. 1980. Sulfur content of plant material: A comparison of methods of oxidation prior to determination. *Commun. Soil Sci. Plant Anal.* 11:257–266.

Soon, Y.K., Y.P. Kalra, and S.A. Abboud. 1996. Comparison of some methods for the determination of total sulfur in plant tissues. *Commun. Soil Sci. Plant Anal.* 27:809–818

Sterrett, S.B., C.B. Smith, M.P. Mascianica, and K.T. Demchak. 1987. Comparison of analytical results from plant analysis laboratories. *Commun. Soil Sci. Plant Anal.* 18:287–299.

Tabatabai, M.A., N.T. Basta, and H.J. Pirela. 1988. Determination of total sulfur in soils and plant materials by ion chromatography. *Commun. Soil Sci. Plant Anal.* 19:1701–1714.

Topper, K. and J. Kotuby-Amacher. 1990. Evaluation of a closed vessel acid digestion method for plant analyses using inductively coupled plasma spectrometry. *Commun. Soil Sci. Plant Anal.* 21:1437–1455.

Watson, M.E. 1981. Interlaboratory comparison in the determination of nutrient concentrations of plant tissue. *Commun. Soil Sci. Plant Anal.* 12:601–618.

Watson, M.E. and R.A. Isaac. 1990. Analytical instruments for soil and plant analysis, pp. 691– 740. In: R.L. Westerman (Ed.), *Soil Testing and Plant Analysis.* Third edition. SSSA Book Series 3. Soil Science Society of America, Madison, WI.

Wolf, B. 1982. A comprehensive system of leaf analyses and its use for diagnosing crop nutrient status. *Commun. Soil Sci. Plant Anal.* 13:1035–1059.

Zhao, F, S.P. McGrath, and A.R. Crosland. 1994. Comparison of three wet digestion methods for the determination of plant sulphur by inductively coupled plasma atomic emission spectroscopy (ICP-AES). *Commun. Soil Sci. Plant Anal.* 25:407–418.

SULFATE-SULFUR DETERMINATION IN PLANT TISSUE

13

C. Grant Kowalenko

INTRODUCTION

The sulfate-sulfur (SO_4-S) content of plants has been used as an indicator of their S nutrient status (Beaton et al., 1968). The basic principle is based on the observation that, although plants take up S as the sulfate (SO_4^{2-}) anion from the soil, the dominant form of S in the plant is reduced S such as S-containing amino acids and related organic compounds (Dijkshoorn et al., 1960; Lakkineni and Abrol, 1994; Mengel and Kirkby, 1987; Schmidt and Jager, 1992). Sulfate is assumed to be transitory in the plant since it is reduced quickly for incorporation into plant components, and will accumulate only when it is in excess to plant requirement. The accumulation of SO_4 in plants, however, is influenced by many factors (e.g., age of plant, type of plant, status of other nutrients) that can sometimes make interpretation of measurements complex.

The measurement of SO_4 in plants has been done in a very wide variety of ways (Beaton et al., 1968). The two basic methods involve extracting the plant sample with water or a chemical solution and measuring the SO_4 in the solution, or measuring the SO_4 directly on the sample using reduction with a hydriodic acid-containing reagent.

Many different extraction solutions have been used, but it has not been determined whether or not the solution extracts all of the SO_4 present in the

1-57444-124-8/98/$0.00+$.50
© 1998 by CRC Press LLC

plant nor whether or not reduced forms of organic S are converted to sulfate during extraction and subsequent quantification. A number of methods have been used to quantify the SO_4 in the extract, all of which have important implications for interpretation. The most common methods of quantification have been the hydriodic acid reduction method and gravimetry, colorimetry, or turbidimetry involving precipitation with barium (Ba). Barium-based methods are assumed to be specific to inorganic SO_4, but are not particularly sensitive and subject to interference from many organic and inorganic compounds. The hydriodic acid method is quite sensitive and free from interferences, but includes organic as well as inorganic SO_4 and is operationally slow. More recently, ion chromatography (Stevens, 1985) and inductively coupled plasma atomic emission spectrometry (Novozamsky et al., 1986) have been developed and are capable of S measurements. Ion chromatography measures the inorganic SO_4 form of S quite specifically, but requires specialized instrumentation. Since the inductively coupled plasma spectrometer measures total S (organic and inorganic, oxidized and reduced), SO_4 must be separated from other forms before quantification to make it S-form specific. Novozamsky et al. (1986) used Ba precipitation for S species separation, making the method similar to other Ba-based methods. To date, there are little data by these newer methods on which to evaluate interpretation of the measurements.

The application of the hydriodic acid reduction method directly on plant material has the advantage of eliminating the extraction step, as well as its relative sensitivity and freedom from interferences. There are a number of studies with this direct measurement that have shown its effectiveness for determining the S nutritional status of plants in a variety situations (Janzen and Bettany, 1984; Millard et al., 1985; Pinkerton and Randall, 1995; Scott et al., 1983, 1984). Although the method includes both organic and inorganic SO_4, it is quite specific to SO_4 (Tabatabai, 1982). Adoption of a bismuth sulfide (BiS) instead of a methylene blue colorimetric determination of the hydrogen sulfide (H_2S) produced by the hydriodic acid reagent (Kowalenko and Lowe, 1972) and redesigning of the digestion-distillation apparatus (Kowalenko, 1985) has reduced the time for an analysis to less than 10 minutes from the 60 to 120 minutes required by the original method proposed by Johnson and Nishita (1952).

EQUIPMENT

1. Reduction/distillation apparatus custom built from readily available glass materials as described by Kowalenko (1985).

2. Heater capable of maintaining a custom-built aluminum heating block to fit the digestion/distillation flask at 110 to 115°C, with a suitable heat shield and a mechanism for lifting/supporting the reduction/distillation apparatus as described by Kowalenko (1985).
3. Laboratory oven for drying plant and aqueous standard samples.
4. Analytical balance with 0.0001 readability.
5. Flowmeter to deliver nitrogen (N_2) gas through digestion/distillation apparatus at 200 mL/min (Kowalenko, 1985) or 30-cm length of capillary tubing (Kowalenko and Lowe, 1972).
6. Test tubes for hydrogen sulfide (H_2S) gas absorption. The size and shape of the tube should be selected to ensure that the N_2 bubbles through at least 4 to 5 cm of absorption solution and to accommodate a specific SO_4-S detection range (5 mL for 0 to 40 µg S/mL and 20 mL for 0 to 200 µg S/mL).
7. Spectrophotometer for measurements of solutions at 400 nm having volumes as small as 7.5 mL.

REAGENTS

1. **Nitrogen Gas Purification Solution:** dissolve 5 to 10 g mercuric chloride ($HgCl_2$) in 100 mL 2% potassium permanganate (K_2MnO_4).
2. **Hydriodic Acid Reducing Solution:** mix four volumes of hydriodic acid (e.g., 57% with 1 to 2.5% hypophosphorus acid preservative), 1 volume of hypophosphorus acid (50%) and 2 volumes of formic acid (90%), then heat for 10 min at 115 to 117°C while bubbling purified N_2 through it during both heating and cooling steps. Volatile gases during heating should be controlled by adequate ventilation, refluxing, and/or use of a gas scrubber (Tabatabai, 1982). Only sufficient reagent for a few days of analyses should be prepared and stored in a brown bottle under refrigeration to minimize degradation that can otherwise occur.
3. **Hydrogen Sulfide Absorption Solution:** 1M sodium hydroxide (NaOH).
4. **Sulfide Colorimetric Solution:** mix bismuth nitrate [$Bi(NO_3)_2 \cdot 5H_2O$] reagent (heat 3.4 g of pentahydrate form in 230 mL glacial acetic acid; filter if not clear) and gelatin solution (heat 30 g in 500 mL water) and make to 1 L with water at room temperature. This solution is stable at room temperature.
5. **Sulfate-Sulfur Solution:** working standards are made by appropriate dilution from a 1,000 mg S/mL by dissolving 5.435 g dried reagent grade potassium sulfate (K_2SO_4) in water and made to 1 L.

6. **Nitrogen (N$_2$) Gas:** purified and of a suitable source for sustained delivery (e.g., compressed).

PROCEDURE

1. Assemble the reduction/distillation apparatus above the heating block in such a way that the digestion/distillation tube can be installed or removed easily, and that, when the digestion/distillation tube is lowered into the block, the N$_2$ outlet is positioned in the absorption solution (5 mL for 0 to 40 µg S/mL and 20 mL for 0 to 200 µg S/mL calibration ranges) to have the nitrogen freely bubble through at least 4 to 5 cm depth.
2. Attach the N$_2$ gas to the apparatus after bubbling through purification solution and delivery rate regulator.
3. Measure plant samples and standards into digestion/distillation tubes. For plant samples, this should be a predetermined weight of relatively finely ground (fineness should suit the weight of sample used) and oven-dried material. The weight of the sample will depend on the anticipated SO$_4$-S content of the plant (Millard et al., 1985; Scott et al., 1983, 1984 for possible plant SO$_4$ contents) and the range of standards (0 to 40 µg S/mL or 0 to 200 µg S/mL) used for calibration. For standard samples, this should be a small volume of aqueous solution that can be completely dried in a short time.
4. Condition the digestion/distillation apparatus with a 200 µg S/mL standard solution (drying this solution is not essential) with heater and N$_2$ gas flow as outlined. This is done by adding 4 mL hydriodic acid reagent to the digestion/distillation tube containing the sample and quickly installing it to the apparatus and lowering apparatus in place for heating and gas absorption. Ensure there is a complete gas seal of the apparatus; a drop of water at the juncture of the digestion/distillation tube and apparatus can be used for sealing.
5. After 10 min of heating and gas flow, remove the absorption solution and add an appropriate volume of the Bismuth Colorimetric Solution (for every two volumes of absorption solution add one volume of bismuth solution) and mix. Proceed to a sample only if this solution is highly colored.
6. Repeat Steps 4 and 5 above with plant and standard samples.
7. Determine the concentration of S in the absorption-colorimetric solution by standard colorimetric procedures.

ANALYSIS

Standard colorimetric procedures with appropriate standard calibration (including reagent blanks) should be used to determine the SO_4 content in plant samples. The content in the plant should be calculated on a unit weight basis by mathematical adjustment for the weight measured into the digestion/distillation tube.

There are no known certified reference materials for plant SO_4 contents. Establishment of in-house reference samples should be considered as one of many methods for quality assurance.

COMMENTS

Safety precautions should be taken when handling the chemicals and reagents. There should be adequate ventilation, particularly when the hydriodic acid reagent is being heated. There is the possibility of producing phosphine if the reducing acid reagent is heated above 120°C.

The original procedure included N_2 gas purification (Johnson and Nishita, 1952). This may be omitted by using high-purity gas, but the purity of the gas could be evaluated by running reagent blanks; this evaluation should be done as often as the uniformity of the gas may change (e.g., check each tank if compressed gas is used).

The time of digestion/distillation will depend on the internal volume of the apparatus and N_2 gas flow rate. Tests should be run on the apparatus with standards to determine the time required for complete digestion and transfer to the absorption solution (Kowalenko, 1985).

Water will influence the hydriodic acid reduction reaction (Kowalenko and Lowe, 1972); therefore, the amount of water in all materials (plant and standard solutions) should be uniform. This can most easily be done by oven drying all standards and samples.

Volumes of the absorption and colorimetric solutions of both the standard and plant samples must be measured very consistently for precise results.

Different methods to measure plant SO_4 (direct hydriodic acid method; extraction followed by barium, hydriodic acid or ion chromatography quantification, etc.) should not be expected to yield precisely the same results; therefore, it would be helpful to document and briefly highlight the method of analysis. The use of a term such as hydriodic-acid-reducible (or HI-) S may be very useful to remind readers on the importance of methodology for interpretation and comparison of data.

The ratio of SO_4-S to total S in plants has been proposed as a good indicator for the plant S nutrient status; however, Scaife and Burns (1986) showed that this approach to interpretation of data should be done with extreme caution. Sulfate-sulfur itself appears to be better, more straightforward and simpler analytically than the SO_4/total S ratio because only one analysis is required.

REFERENCES

Beaton, J.D., G.R. Burns, and J. Platou. 1968. Determination of Sulphur in Soils and Plant Material, pp. 17–19. Technical Bulletin 14. The Sulphur Institute, Washington, D.C.

Dijkshoorn, W., J.E.M. Lampe, and P.F.J. Van Burg. 1960. A method of diagnosing the sulphur nutrition status of herbage. *Plant Soil* 13:227–241.

Janzen, H.H. and J.R. Bettany. 1984. Sulfur nutrition of rapeseed: I. Influence of fertilizer nitrogen and sulfur rates. *Soil Sci. Soc. Am. J.* 48:100–107.

Johnson, C.M. and H. Nishita. 1952. Microestimation of sulfur in plant materials, soils and irrigation waters. *Anal. Chem.* 24:736–742.

Kowalenko, C.G. 1985. A modified apparatus for quick and versatile sulphate-sulphur analysis using hydriodic acid reduction. *Commun. Soil Sci. Plant Anal.* 16:289–300.

Kowalenko, C.G. and L.E. Lowe. 1972. Observations on the bismuth sulfide colorimetric procedure for sulfate analysis in soil. *Commun. Soil Sci. Plant Anal.* 3:79–86.

Lakkineni, K.C. and Y.P. Abrol. 1994. Sulphur requirement of crop plants: Physiological analysis. *Fertilizer News* 39:11–18.

Mengel, K. and E.A. Kirkby. 1987. Principles of Plant Nutrition, pp. 385–402. Fourth edition. International Potash Institute, Bern, Switzerland.

Millard, P., G.S. Sharp, and N.M. Scott. 1985. The effect of sulphur deficiency on the uptake and incorporation of nitrogen in ryegrass. *J. Agric. Sci. (Camb.)* 105:501–504.

Novozamsky, I., R. van Eck, J.J. van der Lee, V.J.G. Houba, and E. Temminghoff. 1986. Determination of total sulphur and extractable sulphate in plant materials by inductively coupled plasma atomic emission spectrometry. *Commun. Soil Sci. Plant Anal.* 17:1147–1157.

Pinkerton, A. and P.J. Randall. 1995. Sulfur requirement during early growth of *Trifolium balansae, Trifolium subterraneum, Medicago murex,* and *Phalaris aquatica. Austr. J. Exp. Agric.* 35:199–208.

Scaife, A. and I.G. Burns. 1986. The sulphate-S/total S ratio in plants as an index of their sulphur status. *Plant Soil* 91:61–71.

Schmidt, A. and K. Jager. 1992. Open questions about sulfur metabolism in plants. *Annu. Rev. Plant Physiol. Mol. Biol.* 43:325–349.

Scott, N.M., P.W. Dyson, J. Ross, and G.S. Sharp. 1984. The effect of sulphur on the yield and chemical composition of winter barley. *J. Agric. Sci. (Camb.)* 103:699–702.

Scott, N.M., M.E. Watson, K.S. Caldwell, and R.H.E. Inkson. 1983. Response of grassland to the application of sulphur at two sites in north-east Scotland. *J. Sci. Food Agric.* 34:357–361.

Stevens, R.J. 1985. Evaluation of the sulphur status of some grasses for silage in Northern Ireland. *J. Agric. Sci. (Camb.)* 105:581–585.

Tabatabai, M.A. 1982. Sulfur, pp. 501–538. In: A.L. Page, R.H. Miller, and D.R. Keeney (Eds.), *Methods of Soil Analysis. Part 2. Chemical and Microbiological Properties.* Second edition. Agronomy Monograph Number 9. American Society of Agronomy, Madison, WI.

DETERMINATION OF CHLORIDE IN PLANT TISSUE

14

Liangxue Liu

INTRODUCTION

Although chloride (Cl) is classified as an essential plant micronutrient (not more than 100 mg Cl/kg for biochemical functions), plants can normally accumulate much higher concentrations in the range of 2,000 to 20,000 mg Cl/kg. The majority of the Cl present in the plant is in the ionic form; therefore, Cl in the plant is quantitatively extracted with water or diluted acid or diluted salts (Gaines et al., 1984).

EQUIPMENT

1. 90-mL glass vial with snap cap.
2. Reciprocating or rotating shaker capable of 250 rpm.
3. Whatman No. 42 filter paper.

REAGENTS

1. Deionized high-resistance water (>18MΩ).
2. **Renex 30, 30% (w/w) Solution:** shake the container of Renex 30 until it

1-57444-124-8/98/$0.00+$.50

appears homogeneous. Transfer an excess of 30 g Renex 30 to a 4-oz. polyethylene dropper bottle. Add the Renex 30 from the polyethylene bottle dropwise to a 250-mL beaker containing 70 mL deionized water while mixing with a magnetic stirrer. Slow addition is necessary to avoid clumping of the Renex 30.

3. **Chloride Color Solution:** dissolve 520 mg mercuric thiocyanate [$Hg(CNS)_2$] in 200 mL methanol in a 1-L flask. Add 600 mL deionized water and mix. Add 63 mL concentrated nitric acid (HNO_3) and mix. Dissolve 40 g ferric nitrate [$Fe(NO_3)_3 \cdot H_2O$]. Dilute to 1 L with deionized water and mix thoroughly. Add 1.0 mL of the Renex 30, 30% Solution and mix well.

4. **System Wash Solution (4.5N sulfuric acid):** slowly add 125 mL concentrated sulfuric acid (H_2SO_4) to about 600 mL deionized water. Mix and allow to cool to room temperature. Dilute to 1 L with deionized water and mix thoroughly.

5. **Chloride Primary Standard** (1,000 mg/L): weigh 2.103 g oven-dry potassium chloride (KCl) and dilute to 1 liter with deionized water. Prepare a series of working standards by diluting aliquots of the primary standard with deionized water.

PROCEDURE

1. Grind dried plant material to a fine powder.
2. Weigh 250 mg of the fine powder into a 90-mL glass vial.
3. Add 25 mL deionized water and cap the vial.
4. Place the vial on a reciprocating shaker and shake at 250 rpm for 30 minutes.
5. Remove from the shaker and allow to stand for 15 minutes.
6. Filter through Whatman No. 42 filter into a plastic vial for Cl analysis.

ANALYSIS

The Cl in the filtrate can be analyzed using the colorimetric method on the TRAACS 800™ AutoAnalyzer (Tel and Heseltine, 1990). In this method, the sample is mixed with the color reagent and dialyzed into the color reagent again. The procedure is based on the release of thiocyanate ions from mercuric thiocyanate by Cl ions in the sample. The liberated thiocyanate reacts with ferric iron to form a red color complex of ferric thiocyanate. The color of the resulting solution is stable and directly proportional to the original Cl concentration. The color complex is measured at 480 nm using a 10-mm flow cell. Nitrite (NO_2), sulfide, cyanide, thiocyanate, bromide, and iodine ions cause interferences when present in sufficient amounts (Fixen et al., 1988).

COMMENTS

1. This procedure is very sensitive. Highly colored plant extracts cause no interference due to dialysis.
2. The water extractable Cl gives a very good estimate of Cl present in the plant.

ACKNOWLEDGMENT

Chapter is dedicated to the late Mr. Dirk Tel for his outstanding contribution to the development of many analytical methods.

REFERENCES

Fixen, P.E., R.H. Gelderman, and J.L. Denning. 1988. Chloride tests, pp. 26–28. In: Dahnke, W.C. (Ed.), *Recommended Chemical Soil Test Procedures for the North Central Region*. North Central Regional Publications No. 221 (revised).

Gaines, T.P., M.B. Parker, and G.L. Gascho. 1984. Automated determination of chlorides in soil and plant tissue by sodium nitrate extraction. *Agron. J.* 76:371–374.

Tel, D.A. and C. Heseltine. 1990. Chloride analyses of soil leachate using the TRAACS 800 Analyzer. *Commun. Soil Sci. Plant Anal.* 21:1689–1693.

EXTRACTABLE CHLORIDE, NITRATE, ORTHOPHOSPHATE, POTASSIUM, AND SULFATE-SULFUR IN PLANT TISSUE: 2% ACETIC ACID EXTRACTION

15

Robert O. Miller

SCOPE AND APPLICATION

The method semiquantifies the concentration of chloride (Cl), nitrate-nitrogen (NO_3-N), orthophosphate-phosphorus (PO_4-P), potassium (K), and sulfate-sulfur (SO_4-S) in plant tissue by extraction with a 2% acetic acid solution. Dilute acetic acid does not quantitatively extract these anions from the tissue. Nitrate is determined spectrophotometrically at 520 nm by the Griess-Ilasvay method (cadmium reduction); K is determined by atomic emission or absorption spectrometry; PO_4-P in the extract is determined spectrophotometrically at 660 nm by reacting with paramolybdate; and Cl is determined by coulometric titration or ion-selective electrode (Watson and Isaac, 1990). The method has been used primarily to determine NO_3-N, K, PO_4-P, SO_4-S, and Cl for assessing plant fertility and Cl status (Johnson and Ulrich, 1959; Chapman and Pratt, 1961). The method can also be used to determine extractable ammonium-nitrogen

(NH$_4$-N). Generally the method detection limit is approximately 10 mg/kg (sample dry basis) and is generally reproducible to within ±10.0 %.

EQUIPMENT

1. Analytical balance, 250-g capacity, resolution ±0.1 mg.
2. Reciprocating mechanical shaker, capable of 180 oscillations per minute.
3. 125-mL extraction vessel with cap and filtration container.
4. Repipette dispenser calibrated to 50.0 ± 0.2 mL.
5. Whatman No. 2V 11-cm filter paper or equivalent highly retentive paper.
6. UV-VIS spectrophotometer, 520 and 660 nm.
7. Atomic emission/absorption spectrometer.
8. Coulometric titrator or chloride ion-selective electrode.

REAGENTS

1. **Acetic Acid Extraction Solution:** dilute 20 mL acetic acid in 50 mL deionized water and dilute to 1.0 L. Care must be taken to use high-purity acetic acid to avoid NO$_3$ and Cl contamination.
2. **Standard Calibration Solutions** (Cl, NO$_3$-N, PO$_4$-P, K, and SO$_4$-S): prepare multiple calibration standards according to specific method and manufacturer's specifications prepared from 1,000 mg/L standard solution and diluted to final volume with 2% acetic acid.

PROCEDURE

1. Weigh 200.0 ± 1.0 mg air-dried plant tissue (see Comments 1 and 2) and place in 125-mL extraction vessel. Include a method blank.
2. Add 50.0 ± 0.2 mL 2% Acetic Acid Extraction Solution and place on reciprocating mechanical shaker for 30 minutes (see Comment 3). Include a method blank.
3. Filter, refilter if filtrate is cloudy (see Comments 4, 5, 6, and 7) and retain for analysis.
4. Analysis
   - **Determination of Cl:** analyze for Cl according to Method S-1.40 or method 407-B listed in "Standard Method for the Analysis of Waste Water" (Franson, 1985). Record concentration in mg Cl/L in extract.

- **Determination of NO_3-N:** analyze using spectrophotometric method 418-6 or 41 8-F for NO_3-N listed in "Standard Method for the Analysis of Waste Water" (Franson, 1985). Record concentration in mg NO_3-N/ L in extract.
- **Determination of PO_4-P:** analyze using spectrophotometric method 424-F in "Standard Method for the Analysis of Waste Water" (Franson, 1985). Record concentration in mg PO_4-P/L in extract.
- **Determination of K:** analyze by atomic emission or atomic absorption spectrometry (Watson and Isaac, 1990). Record concentration in mg K/ L in extract.
- **Determination of SO_4-S:** analyze using turbidimetric method in "Standard Method for the Analysis of Waste Water" (Franson, 1985). Record concentration in mg SO_4-S/L in extract.

CALCULATIONS

Report mg of Cl, NO_3-N, PO_4-P, K, and SO_4-S in sample as to the nearest 10 mg/kg:

$$mg/kg = \frac{(\text{analyte conc. mg/L} - \text{method blank}) \times (250)}{\text{dry matter content (\%)}/100}$$

COMMENTS

1. Plant tissue must be ground to pass 40-mesh screen (<0.40-mm) in order to ensure adequate homogeneity.
2. Sample mass may be adjusted in accordance with expected analyte concentrations. For materials containing <500 mg NO_3-N/kg, increase sample size to 500 mg.
3. Check repipette dispensing volume, calibrate using an analytical balance.
4. Check filter paper supply for possible contamination of analytes. If significant contamination is found (>10 mg/kg on a sample basis), rinse filter paper with acetic acid extraction solution or filter extract with serum separator tubes.
5. Acetic acid extracts may be stored for several days, if stored at 4°C and/or with 100 µL of toluene or thymol.
6. Extracts may be retained for analysis of total K, NH_4-N, and SO_4-S.
7. Samples having Cl, NO_3-N, PO_4-P, K, and SO_4-S concentrations exceeding the highest standard will require dilution and reanalysis.

REFERENCES

Chapman, H.D. and P.F. Pratt. 1961. Methods of Analysis for Soils, Plants and Waters. Priced Publication 4034. Division of Agriculture Sciences, University of California, Berkeley.

Franson, A.H. (Ed.). 1985. *Standard Method of the Examination of Waste Water,* pp. 265–297. Sixteenth edition. American Water Works Association, and Water Pollution Control Federation. American Public Health Association, Washington, D.C.

Johnson, C.M. and A. Ulrich. 1959. *Analytical Methods for Use in Plant Analysis,* pp. 26–78. Bulletin 766. Agricultural Experiment Station, University of California, Berkeley.

Watson, M.E. and R.A. Isaac. 1990. Analytical instruments for soil and plant analysis, pp. 691–740. In: R.L. Westerman (Ed.), *Soil Testing and Plant Analysis.* SSSA Book Series 3. Soil Science Society of America, Madison, WI.

TISSUE TESTING KITS AND PROCEDURES FOR NUTRIENT ELEMENT ASSESSMENT IN PLANT TISSUE

16

J. Benton Jones, Jr. and Denton Slovacek

INTRODUCTION

Tissue testing kits are designed for use in the field, providing an on-the-spot assessment of the nutrient element status of a plant. Normally, the tests are performed using apparatus that has been specifically designed for in-field use, although some kits may have laboratory application. Nutrient element determinations are limited to those elements [mainly nitrogen (N) as nitrate (NO_3), phosphorus (P), and potassium (K)] or parameters that can be measured with procedures that are easy to conduct using one or a combination of reagents, chemically treated test papers, vials, and/or simple testing apparatus. Test procedures may either rely entirely on electronic devices and instruments or they may be totally non-instrumental.

TISSUE TESTING KITS

A tissue testing kit is an analytical system that contains all the necessary reagents, apparatus, detection system, and documentation packaged as a portable

1-57444-124-8/98/$0.00+$.50
© 1998 by CRC Press LLC

kit. A well designed and manufactured tissue testing kit should be safe to transport and use and able to provide the user with reasonably accurate analytical results that can be interpreted.

TISSUE TESTING KIT TECHNOLOGY: AN OVERVIEW

Tissue testing kits are often viewed by some as marginal analytical systems, convenient to use but not sufficiently accurate for serious data accumulation. Therefore, some test kits have often been relegated for use as *educational tools* providing the user with an interesting hands-on experience in conducting an analysis, but not suitable for making an accurate assessment of the plant's nutritional status.

This limited view of kit technology, in large part, has been due to those test kits that are qualitative or semiquantitative in nature, providing no proof of accuracy, and consisting of apparatus that is often of poor quality and accuracy. Documentation is often limited and of poor quality. Included reagents may have been packaged in such a way that they are unstable, and in extreme cases, unsafe to transport and use.

Present-day test kits, designed and manufactured by competent companies, have overcome these problems and they often exceed the expectations of the user. They provide high-quality reagents that are safe, stable, and pure. A well-designed test kit will include complete documentation that is well written and concise with additional information provided, such as material safety data sheets, re-order information, and information that is relevant to the analytical techniques needed to perform the tests.

A test kit should include all the required reagents and test items in sufficient quality needed to conduct more than 25 assays.

Interpretation of a tissue test result may require considerable skill and experience on the part of the user, which can only be obtained by repeated testing and evaluation on plants of known nutrient element status. Therefore, tissue testing may be viewed as a procedure that has considerable *art* or *acquired skill* associated with its use even although detailed interpretive information may be provided with the kit (Armstrong, 1984).

DETECTION SYSTEM

A most important part of the test kit is the detection system by which the user determines the value needed for the calculation of results, a functional detection system designed for durability and portability, of sufficient accuracy and capable of detection within the normal analyte concentration range of the prepared sample. Finally, the detection system should be simple to use, transport, and store. Being a portable analytical system, there may be limitations on the type of detection system most suited for a particular analyte, although most detection systems are based on either colorimetric and/or titrimetric technology.

There are two types of colorimetric detection: (1) visual comparison of color, and (2) photometric measurement of color. Visual comparison is accomplished at low cost, although even the best visual comparator systems are limited by visual acuity, lighting, and the stability of the standard color used as the comparator with a normal variance of ±10% around the expected true value. For example, a standard sample containing 1.0 mg P/L can have a reading spread of 0.9 to 1.1 mg P/L, depending on the color comparing acuity of the user. However, this variation will be magnified by the dilution factor, which in turn may influence the interpretation of the obtained analyte value. On the other hand, a well-designed spectrophotometer instrument has a variance of ±1% around the true value, but its inclusion in the testing kit adds considerable cost to the analytical system.

Titrimetric determinations are made through a drop-count titration or by using a calibrated dispensing apparatus, such as a burette, syringe, etc. In a drop-count titration, the user counts the number of drops of titrant required to reach the end-point (normally 10 to 20 drops). Resolution of detection is normally plus or minus 1 drop; therefore, the percent error is relative to the number of drops.

Calibrated titration devices add to the precision of the method, but also to the cost of the kit. Therefore, the user must decide between the precision of the method versus the cost of the kit.

Other types of detection systems are available, potentiometric devices, such as pH and ion-specific electrodes and conductivity probes, devices that require knowledge as to their proper use, sensitivity to temperature, calibration, and storage requirements between uses, which add to the complexity of operation for the user.

ENZYMATIC REACTIONS

Biologically based tests have been used to assess the elemental status of the plant, primarily for the micronutrients, iron in particular. One test for iron (Fe) uses the activity of peroxidase, which in the presence of Fe and a reactive solution will result in the development of a blue color; the speed of development and color intensity is used as a measure of the presence of *active* Fe. A collected plant leaf blade disk is floated on the reactive solution, and if *active* Fe is present, a dark blue ring will form around the edge of leaf disk. No blue ring formation around the leaf disk would indicate an Fe-deficient plant (Bar-Akiva, 1984; Bar-Akiva et al., 1978).

PLANT TISSUE TESTING KIT TECHNOLOGY

Tissue testing kits for plant nutrient element assessment are based on the determination of extractable non-assimilated elements (Krantz et al., 1948; Wickstrom, 1967; Jones, 1994a,b; 1997). Therefore, the interpretative data based on total plant elemental content (generally referred to as plant analysis) would not applicable to the interpretation of a tissue- or sap-assay result. There are some important exceptions, such as petiole analysis for nitrate-nitrogen (NO_3-N) in cotton, potato, and sugar beet (Ludwick, 1990).

The plant part to be selected for assay is normally conductive tissue, such as petioles, leaf mid-ribs, stem, and plant stalks. The procedure is to squeeze an aliquot of sap from the tissue on to a test paper or strip (Scaife and Stevens, 1983), or into water or a reactive solution, or to slightly macerate the selected tissue and then place it in water and/or an extraction solution; the test is then performed on the obtained extractant.

Interpretation of a tissue test result is much more subject to sampling variability than that for a total plant elemental analysis. In addition, factors such as the time of day, weather conditions, water stress, and ionic balance of absorbed but unassimilated elements can result in a misinterpretation of the analytical tissue test results.

PLANT CHECK TISSUE TESTING KIT

This tissue testing kit, shown in Figure 16.1, uses a single test paper on which phosphorus (P), nitrate-nitrogen (NO_3-N), and potassium (K) determinations are made on extracted plant sap. The tests for P and NO_3-N are qualitative, while that for K is semiquantitative. Selecting conductive tissue, usually a petiole or leaf mid-rib, an aliquot of sap is squeezed onto a designated portion of the test paper and reagents applied to development the color. The use of this test kit is illustrated in the video by Jones (1994b).

For P, the molybdenum-blue method is employed, the speed and intensity of blue color development (no color: very low; light blue: low; medium blue: medium; and intense blue: high) is used to evaluate the P status of the plant.

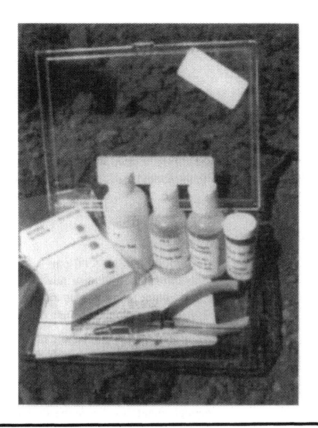

FIGURE 16.1 PLANT CHECK Tissue Testing Kit.

For NO_3-N, the sulfanic acid/naphthylamine pink-red color reaction is used, the speed and intensity of color development (no color or white: very low; pink: low; light red: medium; and cherry red: high) is used to evaluate the N status of the plant. For K, the formation of an orange pigment on one or all three (one: low; two: medium; or all three: high) dots on the test paper is used as an indicator of the K status in the plant. No orange pigmentation on any of the dots means very low K.

Setting-up the PLANT CHECK Tissue Testing Kit
(use these testing materials in the year of purchase only)

Your kit contains the following:

- Potassium test papers
- Nitrate powder
- Extracting pliers
- P-2 extra capsules
- PK-1 bottle
- P-2 bottle
- PK-1 concentrate supply

You need to obtain distilled water from a drug store, bottling company, or a laboratory. Add distilled water to the shoulder of each the PK-1 and P-2 bottles. Your kit is now ready to use.

Plant Sampling

When	From 8:00 a.m. to 5:00 p.m. Not immediately after a rain. From young plant to near maturity. Not during drought or other stress conditions.
Part of Plant (see Sampling Instructions)	Mid-rib or petiole of recently matured leaf (avoid old at bottom, young at top). Stalk, midway up plant.
Part of Field	Avoid bad spots except for comparison. Test enough plants to determine definite pattern of levels.

Testing Procedures

Nitrate Test

- **On plant**—Split the stalk, mid-rib, or petiole. Add small amount (match-head size) of nitrate powder to cut tissue and work into sap with knife blade. Wait 5 minutes for final reading.
- **On test paper**—Fold the corner of paper and place a section of plant tissue and the same amount of nitrate powder as given above in the fold. Squeeze fold until sap contacts powder. Wait 5 minutes for final reading.
- **In vials**—Mash the equivalent of 1/8 teaspoon of tissue with pliers, place in vial containing 5 mL of water. Stir one minute, add pea-size portion of the nitrate powder. Shake and allow 5 minutes for reaction.
- **Nitrate-Nitrogen Stalk Test**—For a crop plant such as corn, cut a 3- to 4-inch section at the base of the stalk. Cut the stalk section in half, place some nitrate powder on the open cut, put the two halves together, and move them

in order to mix the powder with the exposed stalk cut. In a few minutes, open the two halves. If nitrate (NO_3) is present, a pink to red color will indicate its presence, the intensity of the color indicating the concentration of NO_3-N.

- **Reading**—No color or white: very low, pink: low, light red: medium, cherry red: high.

Phosphorus Test

- **On test paper**—Squeeze small amount of sap from cut end of tissue onto filter paper. Add small drop of PK-1, then a large drop of P-2.
- **In vials**—Place 5 mL of PK-1 solution in the vial. Mash and stir tissue as for the nitrate test. Stir for 1 minute, add a drop of P-2 or stir with a tin rod.
- **Phosphorus stalk test**—For a crop plant such as corn, cut a 3- to 4-inch section at the base of the stalk. Cut the stalk section in half, and place two to three drops of Phosphorus Reagent No. 1 on the open cut followed by two to three drops of Phosphorus Reagent No. 2. Put the two halves together, moving them in order to mix the added reagents with the exposed stalk cut. In a few minutes, open the two halves. If phosphorus (P) is present at a minimum concentration, the blue color will indicate its presence, the intensity of the blue color indicating the concentration of P.
- **Reading**—No color: very low, light blue: low, medium blue: medium, intense blue: high.

Potassium Test

On test paper—Squeeze small amount of sap on each of the orange test dots. Allow 30 to 40 seconds for reaction to take place. Then wash each dot with PK-1 (not an excess) to removal orange color that will go.

- **Reading**—Orange left on sapspot, all 3 dots: high, orange left on medium and low sap spots: medium, orange left on low sap spot: low, no orange color left: very low.

Precaution

Summer heat, light, and contamination can cause chemicals to deteriorate. Check them as follows:

- Nitrate powder should be white in color, not grey or pink.
- Phosphorus solution—Saliva is high in phosphate (PO_4). Check solutions by moistening filter paper with tongue. Run phosphorus test on this and on

a blank spot (no phosphorus). Either no reaction where phosphate is present or color reaction where it is not, calls for a change of solutions. Simply wash bottles, add PK-1 concentrate from supply bottle to bottom mark and distilled water to shoulder. For new P-2 solution, place contents of one capsule in P-2 bottle and add distilled water to fill the bottle.
- Potassium paper dots should be bright orange in color. When washed with PK-1, the dots should be yellow, not brownish.

PREPARATION OF REAGENTS FOR CONDUCTING TISSUE TESTS USING A FILTER PAPER (Source: Syltie et al., 1972)

Nitrate-Nitrogen (NO₃-N)

Reagent 100 grams dry barium sulfate ($BaSO_4$), 10 grams manganese sulfate ($MnSO_4 \cdot H_2O$), 2 grams of finely powdered zinc (Zn), 75 grams citric acid, 4 grams sulfanic acid, and 2 grams of ∂-naphthylamine are finely ground as separate portions with a mortar and pestle, then thoroughly mixed and stored in a blackened container.

Reaction: Any degree of red color produced on reaction with plant sap indicates the presence of nitrate (NO_3).

Phosphorus (P)

- **Solution A**—10 grams ammonium molybdate [$(NH_4)_6Mo_7O_{24} \cdot 4H_2O$] are dissolved in 85 mL water.
- **Solution B**—Mix 16 mL water with 170 mL concentrated hydrochloric acid (HCl).
- **Concentrated solution**—Mix solutions A and B and add 2 grams boric acid (H_3BO_3) per 50 mL of the mixed solution.
- **Working solution**—Dilute the concentrated solution 10 times with water.
- **Reduction suspension**—Place tin chloride ($SnCl_2 \cdot 2H_2O$) in a small dropping bottle and add water.

Potassium (K)

- **Solution A**—Add 0.6 grams dipicrylamine (2,2′,4,4′,6,6′hexanitrodiphenylamine) and 0.6 grams of sodium carbonate (Na_2CO_3) to 25 mL water and boil for 10 minutes.
- **Solution B**—Dilute 8 mL of solution A to 25 mL with water.

- **Solution C**—Dilute 10 mL of solution B to 15 mL with water.
- **Preparation of filter paper**—Three separate 8-mm diameter spots, one each from solutions A, B, and C, are placed on a filter paper and allowed to dry.

SPECIFIC ION NITRATE METER

A relatively quick method for determining the nitrate-nitrogen (NO_3-N) level in petiole cell sap is done with the use of a specific ion meter, such as the Cardy meter as shown in Figure 16.2. The use of the Cardy meter is illustrated in a video by Jones (1994b). The procedure is as follows:

- Collect a representative sample of leaf or petiole tissue.
- Using a sap press (garlic press), squeeze an aliquot of sap onto a clean smooth plastic surface.
- Transfer an aliquot of the sap directly onto the meter sensor and read the NO_3-N concentration.

Using a reference source relating nitrate-nitrogen (NO_3-N) content with nitrogen (N) plant status (Ludwick, 1990), compare the meter reading obtained with the reference to determine if the concentration found is within the suffi-

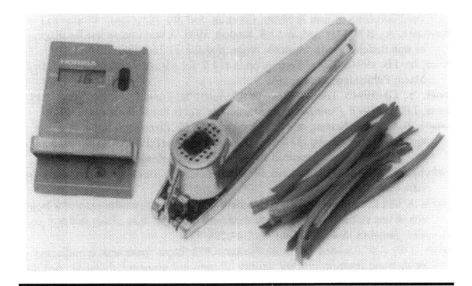

FIGURE 16.2 Cardy Nitrate-Specific Ion Meter.

ciency range for the plant part being tested, the type of crop, and stage of crop development.

INSTRUMENTATION AND KIT SOURCES

The major suppliers of test kits for conducting tissue tests:

- HACH Company, P.O. Box 389, Loveland, CO 80539
- LaMotte Chemical, P.O. Box 329, Chestertown, MD 21620
- Spectrum Technologies, 23839 West Andrew Road, Plainfield, IL 60544

For specific instruments used in tissue testing procedures, such as specific ion meters and chlorophyll meters:

- Spectrum Technologies, 23839 West Andrew Road, Plainfield, IL 60544

REFERENCES

Armstrong, D. (Ed.). 1984. *The Diagnostic Approach to Maximum Economic Yields.* Potash and Phosphate Institute, Norcross, GA.

Bar-Akiva, A. 1984. Substitutes for benzidine as N-donors in the peroxidase assay for rapid diagnosis of iron in plants. *Commun. Soil Sci. Plant Anal.* 15:929–934.

Bar-Akiva, A., D.N. Maynard, and J.E. English. 1978. A rapid tissue test for diagnosis of iron deficiencies in vegetable crops. *HortSci.* 13:284–285.

Jones, Jr., J.B. 1994a. Tissue Testing, pp. 5.1–5.9. In: *Plant Nutrition Manual.* Micro-Macro Publishing, Athens, GA.

Jones, Jr., J.B. 1994b. Tissue Testing (VHS video). St. Lucie Press, Boca Raton, FL.

Jones, J.B., Jr. 1997: *Plant Nutrition Manual.* St. Lucie Press, Boca Raton, FL.

Krantz, B.A., W.L. Nelson, and L.F. Burkhart. 1948. Plant-tissue tests as a tool in agronomic research, pp. 137–156. In: H.B. Kitchen (Ed.), *Diagnostic Techniques for Soils and Crops.* American Potash Institute, Washington, DC.

Ludwick, A.E. (Ed.). 1990. *Western Fertilizer Handbook.* Horticultural Edition. Interstate Publishers, Danville, IL.

Scaife, A. and K.L. Stevens. 1983. Monitoring sap nitrate in vegetable crops: Comparison of test strips with electrode methods and effects of time of day and leaf position. *Commun. Soil Sci. Plant Anal.* 14:761–771.

Syltie, P.W., S.W. Melsted, and W.M. Walker. 1972. Rapid tissue tests as indication of yield, plant composition, and fertility of corn and soybeans. *Commun. Soil Sci. Plant Anal.* 3:37–49.

Wickstrom, G.A. 1967. Use of tissue testing in field diagnosis, pp. 109–112. In: G.W. Hardy et al. (Eds.), *Soil Testing and Plant Analysis.* Part 1. SSSA Special Publication No. 2. Soil Science Society of America, Madison, WI.

CHLOROPHYLL METER METHOD FOR ESTIMATING NITROGEN CONTENT IN PLANT TISSUE

17

James S. Schepers, Tracy M. Blackmer,
and Dennis D. Francis

INTRODUCTION

The strong positive relationship between leaf chlorophyll content and leaf nitrogen (N) concentration is the basis used for predicting crop N status. Measurement of leaf N concentration is more traditional than chlorophyll content for evaluating N management practices. However, the difficulty and expense of plant sample collection, preparation, and analysis makes direct determination a cumbersome process. Historically, determination for total N and chlorophyll involved destructive sampling and extensive processing. Chlorophyll meters permit a rapid and non-destructive determination of leaf chlorophyll content by measuring leaf transmittance.

Because the relationship between leaf chlorophyll content and N concentration is not universal for all crops or across cultivars, it is difficult to calibrate chlorophyll meters directly in terms of N concentration (Schepers et al., 1992a,b). In spite of these limitations, chlorophyll meters provide a good comparison of N status between plants of a given cultivar (Wood et al., 1992). Chlorophyll meters also have a unique application in cases where crops continue to take up N without proportional increases in plant growth (i.e., luxury consumption).

1-57444-124-8/98/$0.00+$.50

This occurs because leaf chlorophyll content tends to reach a plateau when other factors become growth limiting, even though crop N uptake continues (Schepers et al., 1992b). This unique aspect of chlorophyll production forms the basis for its use as a tool for estimating crop N status.

EQUIPMENT

1. Minolta SPAD-502 chlorophyll meter as shown in Figure 17.1.
2. An integrating sphere coupled with a spectroradiometer can provide similar data as the above chlorophyll meter. This combination of instruments is

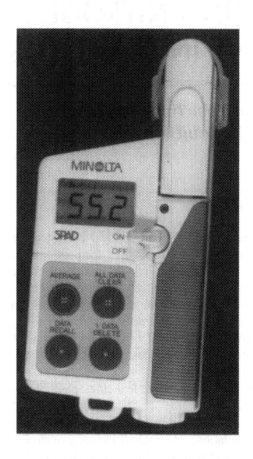

FIGURE 17.1 Minolta SPAD-502 chlorophyll meter.

much more costly than a chlorophyll meter, but more versatile, because it can measure transmittance through the leaf blade or reflectance from the leaf surface over a wide range of wavelengths.

CHLOROPHYLL METER CHARACTERISTICS

Measuring leaf chlorophyll content with a portable meter is a relatively new technology (Benedict and Swidler, 1961) that was introduced by the Minolta Corporation for use in rice production. Since then, application has been extended to a variety of crops. The Minolta SPAD-502 is the only commercially available portable chlorophyll meter at this time.

The meter operates by clamping the forceps-like sensor onto a leaf blade that creates a closed chamber around the area to be measured. A rubber boot seals out external light as the device closes over the leaf blade, which triggers the activation of an internal light source. Radiation not absorbed during photosynthesis is either reflected from the leaf surface or transmitted through the leaf and measured at two dominant wavelengths. The primary wavelength measured is 650 nm (red light), which is sensitive to leaf chlorophyll activity. Meter operation is based on the inverse relationship between absorbed radiation in the 650-nm region of the spectra and that transmitted through the leaf. The second sensor measures the amount of light transmitted at 940 nm (near infrared). This waveband is not affected by leaf chlorophyll content and provides an internal meter calibration. A description of signal processing within the meter is not available. Sensors within the chlorophyll meter (2 × 3 mm) cover a relatively small proportion of the total area of a leaf blade. A consistent sampling protocol is important to increase reliability of the data because of plant-to-plant differences, age differences between leaves, and variation within positions on a leaf. As such, sampling should involve plants having the same growth stage, leaves with comparable age, and a consistent position on leaves (Peterson et al., 1993). Multiple, consistent observations help minimize variability commonly observed with meter readings. Chlorophyll meters permit storage of 30 individual meter readings, reviewing of stored values and elimination of atypical readings, and an averaging function.

PROCEDURE

The sampling strategy for making chlorophyll meter readings should be tailored for each crop and leaf type. Growth stage, relative age of leaves sampled, and position within a leaf should be consistent within a study. The small size of the detectors allows the meter to be used on fine-bladed grasses (i.e., turf, wheat, rice). Sampling strategies have not been published for many crops or applications.

The most extensive application of chlorophyll meter technology has been with corn. Peterson et al. (1993) recommended sampling the uppermost fully expanded leaf of corn (visible collar) until the ear leaf can be identified. They suggest sampling an individual leaf midway between the base and tip and midway between the mid-rib and margin. Other sampling positions on the plant or within a leaf may be appropriate for some applications. Care should be taken not to sample atypical plants within a population; young leaves with non-uniform color; diseased or damaged areas on a leaf; plants adjacent to a missing plant, or plants among multiple plants growing at one location (Blackmer et al., 1993).

When becoming familiar with chlorophyll meter operation and data handling, individuals should consider obtaining an average of 30 readings from representative plants as described above. We recommend sampling across the width of typical field equipment to ensure the sampled area is not affected by fertilizer application, compaction, planter variation, etc. Similar procedures should be followed when sampling comparable plants (same hybrid or variety, planting date, leaf age, sampling position on leaves, etc.) from an area known to have adequate N. It is important that availability of other nutrients be similar when comparing meter readings from adjacent areas.

ANALYSIS

Interpreting chlorophyll meter data is not necessarily obvious because meter readings vary between types of crop, cultivars within a crop, stages of growth, and climatic conditions preceding the measurement. Cultural practices such as type of N fertilizer (Schepers et al., 1992a) and cropping sequence (Schepers et al., 1995) make it difficult to establish a critical level for meter readings. Within a specific crop and stage of growth, it may be possible to establish a threshold value that corresponds to a given situation. Attempts to calibrate a chlorophyll meter usually involve correlations with another type of chemical analysis (i.e., total N, sap nitrate, yield). Non-linear relationships usually result because many crops tend to reach a plateau in chlorophyll content even though continued N uptake may result in luxury consumption. Chlorophyll meter readings tend to be stable within a day, but diurnal fluctuations with other types of measurements (i.e., nitrate, water content) can make correlations difficult to interpret.

Chlorophyll meter readings are affected by crop cultivar, stage of growth, cropping history, and nutrients other than N. Therefore, the easiest way to interpret chlorophyll meter data is to hold those values constant and compare mean readings from areas differing only in the amount of N supplied to the plant. One approach is to reference all meter readings to those from plants where N is not limiting. This calculation results in what has been termed an N Sufficiency Index (NSI):

$$\text{NSI} = \frac{\text{(average meter reading from unknown area)}}{\text{(average meter reading from a comparable area with adequate N)}}$$

Calculation of NSI will result in a value of 1.0 if meter readings are the same for the adequately fertilized reference area and the area in question. As plants become N deficient, meter readings will decline relative to plants from the adequately fertilized reference and the NSI values will be <1.0. The advantage of the NSI concept when interpreting chlorophyll meter data is that it allows comparisons between sampling dates, cultivars, and cropping systems.

The NSI approach has successfully been used to schedule fertigation of corn (Blackmer and Schepers, 1995). The weakness of the index approach is that it requires comparing the same variety, etc. as noted above. It also assumes a uniform level of other nutrients and that interactions between N level and other nutrients are minimal. Interactions between crop water status and leaf chlorophyll content have been documented in a greenhouse study (Schepers et al., 1996), but an N by water interaction is not likely under field conditions with fairly uniform precipitation or irrigation conditions.

COMMENTS

Application of chlorophyll meter technology varies with crop grown and how the N status information is used to make management decisions. Chlorophyll meters have their greatest sensitivity in the deficient to adequate range of N nutrition. Beyond the adequate range, however, the meter does not detect excessive N available to a crop. The strength of the chlorophyll meter is its ability to measure a relative difference in crop N status and to detect the onset of an N stress before it becomes visible.

Managing N availability for crops like corn and sorghum is relatively simple because a slight excess does not adversely affect yield or grain quality. Excess N availability to other crops like cotton, sugar beet, wheat, and barley can adversely affect plant health, yield, quality, and value of the final product.

Therefore, application of chlorophyll meter technology depends on the type of crop being monitored and the reason for managing N availability. Chlorophyll meter readings not only respond to crop N status, but can also be affected by crop water status and availability of other nutrients. Water stress generally decreases chlorophyll meter readings (Schepers et al., 1996). Climatic conditions that affect biological processes, such as mineralization and root development, can also affect nutrient availability. While these potential problems can confound interpretation of chlorophyll meter data, such situations also present unique applications for the meters. One such application of this technology may be to monitor the dynamics of N cycling in soil and compare relative differences in average meter readings (Schepers and Meisinger, 1993).

Another approach for interpreting chlorophyll meter data involves calculating the difference in meter readings between a non-limiting N situation and a possible N-limiting situation. The weakness of the difference approach is that meter readings from a non-limiting situation are likely to change during the growing season. Therefore, the difference between comparison readings may depend on the stage of growth, crop cultivar, cropping history for the field, etc. The NSI approach makes it convenient to compare values between fields, cultivars, growth stages, etc.

To date, researchers have not been able to use absolute chlorophyll meter readings to indicate crop N status or predict crop N requirements for the remainder of the growing season. This is because meter readings are confounded by factors including hybrid or cultivar, plant age, and environmental factors preceding the measurement. If these factors can be eliminated or minimized and generalized interpretations are acceptable, then it may be possible to use absolute meter readings in a meaningful way. One such proposed application for corn is a pre-sidedress critical value to indicate the need for additional N fertilizer (Piekielek and Fox, 1992). Predicting how much fertilizer N to apply is difficult using a chlorophyll meter because, like many other tissue testing procedures, readings represent a point-in-time measurement. Meter readings simply are not able to provide objective information about the N-supplying capacity of soil unless a great deal of other information is known about climatic conditions and soil characteristics. Another application for the chlorophyll meter may be a late-season threshold value to indicate the availability of excess N at maturity. One final way to interpret chlorophyll meter data that has not received much attention is to compare readings from different positions within the same plant. To be meaningful, this approach requires extensive knowledge of plant growth characteristics and makes it necessary to record individual meter readings from the two positions. This approach has merit for crops where plant-to-plant variability is high, but where either an absolute or relative difference between leaf positions within a given plant provides meaningful information. Determining when to harvest alfalfa for maximum protein content may be one application for this approach.

REFERENCES

Benedict, H.M. and R. Swidler. 1961. Nondestructive methods for estimating chlorophyll content of leaves. *Science* 133:2015–2016.

Blackmer, T.M. and J.S. Schepers. 1995. Use of a chlorophyll meter to monitor N status and schedule fertigation of corn. *J. Prod. Agric.* 8:56–60.

Blackmer, T.M., J.S. Schepers, and M.F. Vigil. 1993. Chlorophyll meter readings in corn as affected by plant spacing. *Commun. Soil Sci. Plant Anal.* 24:2507–2516.

Peterson, T.A., T.M. Blackmer, D.D. Francis, and J.S. Schepers. 1993. *Using a Chlorophyll Meter to Improve N Management.* Nebguide G93-1171A. Cooperative Extension Service, University of Nebraska, Lincoln.

Piekielek, W.P. and R.H. Fox. 1992. Use of a chlorophyll meter to predict sidedress nitrogen requirements for maize. *Agron. J.* 84:59–65.

Schepers, J.S. and J.J. Meisinger. 1993. Field indicators of nitrogen mineralization, pp. 31–47. In: C.W. Rice and J.L. Havlin (Eds.), *Soil Testing: Prospects for Improving Nutrient Recommendations.* SSSA Special Publication No. 40. Soil Science Society of America, Madison, WI.

Schepers, J.S., D.D. Francis, M. Vigil, and F.E. Below. 1992a. Comparison of corn leaf nitrogen concentration and chlorophyll meter readings. *Commun. Soil Plant Anal.* 23:2173–2187.

Schepers, J.S., T.M. Blackmer, and D.D. Francis. 1992b. Predicting N fertilizer needs for corn in humid regions: Using chlorophyll meters, pp. 105–114. In: B.R. Bock and K.R. Kelley (Eds.), *Predicting Fertilizer Needs for Corn in Humid Regions.* Bulletin Y-226. National Fertilizer and Environmental Research Center, Tennessee Valley Authority, Muscle Shoals, AL.

Schepers, J.S., G.E. Varvel, and D.G. Watts, D.G. 1995. Nitrogen and water management strategies to reduce nitrate leaching under irrigated maize. *J. Contam. Hydrol.* 20:227–239.

Schepers, J.S., T.M. Blackmer, W.W. Wilhelm, and M. Resende. 1996. Transmittance and reflectance measurements of corn leaves from plants with different nitrogen and water supply. *J. Plant Physiol.* 148:523–529.

Wood, C.W., D.W. Reeves, R.R. Duffield, and K.L. Edmisten. 1992. Field chlorophyll measurements for evaluation of corn nitrogen status. *J. Plant Nutr.* 15:487–500.

ANALYTICAL INSTRUMENTS FOR THE DETERMINATION OF ELEMENTS IN PLANT TISSUE

18

Maurice E. Watson

INTRODUCTION

This chapter acquaints the reader with the types of principal instruments that are typically used for the routine inorganic elemental analysis of plant tissue. Instruments, other than the ones mentioned in this chapter, may be used for conducting a plant analysis, but are not generally considered common to plant analysis laboratories or are used for special, infrequent analysis. Design and operating criteria of a specific instrument model may be different than what is described by the author. This chapter considers the main principles used by the instruments and differentiates those of similar purpose. The advantages and disadvantages of each instrument will be reviewed. Most of the instruments make use of electromagnetic radiation energy. Computers are an integral part of modern laboratory instruments and are used for data capture and storage, setting instrument parameters, and controlling the operation of the instrument. Watson and Isaac (1990), Jones and Case (1990), and Markert (1995), as well as in the book edited by Walsh (1971), have provided a comprehensive review of instruments used for conducting a plant tissue analysis.

1-57444-124-8/98/$0.00+$.50
© 1998 by CRC Press LLC

ATOMIC ABSORPTION SPECTROMETRY

Principle of Operation and Use

Walsh (1955), Baker and Suhr (1982), Tsalev (1984), and Ure (1991) have discussed the principles and usefulness of atomic absorption spectrometry (AAS), and Isaac and Johnson (1975) conducted a collaborative study of the use of AAS for plant tissue analysis.

Atomic absorption spectrometry makes use of the principle that atoms at ground-state energy status can absorb electromagnetic radiation when radiation of appropriate wavelengths is focused on them. The main components of an AAS instrument are: (1) hollow cathode or electrodeless discharge lamps containing the element of interest, (2) nebulizer and burner system, and (3) radiant-energy detector system. The two types of hollow cathode lamps are the single element and the multielement. The element in the lamp is excited by electrical current causing radiation to be emitted from the lamp. The wavelengths of emitted radiation are characteristic for the element contained in the lamp. Electrodeless discharge lamps (EDL) require a microwave power supply to provide greater light output and longer life than hollow cathode lamps. For arsenic (As) and selenium (Se), EDLs provide improved sensitivity and lower detection limits. Some AAS have a turret to hold multiple lamps.

For flame AAS instruments, the sample solution is aspirated into a premixing chamber containing a flow spoiler where large droplets drain away, causing the fine droplets to enter the flame through the burner head. The flame produces ground-state atoms of the element that absorb the light energy from the hollow cathode lamp. Elements that can be measured with this principle must produce sufficient ground-state atoms at the temperature of the flame. Air-acetylene flame is used for most elements. For refractory elements such as aluminum (Al), a higher temperature flame, such as nitrous oxide-acetylene is used.

The light from the flame strikes a monochromator, which isolates the wavelength of interest. Adjustments can be made to allow isolation of wavelengths in the ultraviolet (UV) or visible (VIS) regions of the electromagnetic spectrum. A photomultiplier tube is used to convert the light energy to electrical energy.

Single beam and double beam instruments are the two major optical designs. Double beam AAS instruments are more stable than the single beam instruments. Modern AAS instruments are designed to operate in both absorption and emission modes. The emission mode is usually more stable for the measurement of easily ionizable elements, such as potassium (K) and sodium

(Na), than is the absorption mode. A hollow cathode lamp is not needed when the instrument is used in the emission mode.

In modern instruments, a microcomputer controls the operational parameter settings. Correction for interfering background radiation is often necessary. The Zeeman and Smith-Hjefie are examples of background correction techniques.

Excitation Sources

The air-acetylene flame is the most common excitation source used in AAS. Examples of other excitation sources are acetylene-nitrous oxide, and acetylene-hydrogen gases. Acetylene-nitrous oxide flames are hotter than the air-acetylene flame. Graphite furnace is a flameless excitation source. Hydride generators attached to AAS instruments are used for the determination of very low concentrations of As and Se.

Elements in plant tissue that are determined with air-acetylene flame are K, calcium (Ca), magnesium (Mg), Na, iron (Fe), manganese (Mn), copper (Cu), and zinc (Zn). Other elements such as cobalt (Co) can be determined with flame excitation source, but often require preconcentration before measurement. The heavy metals, nickel (Ni), chromium (Cr), cadmium (Cd), and lead (Pb), can be determined with flame AAS if their concentrations are high enough. For very low concentrations of these heavy metals in plant tissue, graphite furnace AAS is usually used. Flame and flameless AAS have been compared by Rains and Menis (1976).

Advantages

The primary advantages of AAS are: (1) highly specific for an element; (2) minimum spectral interference, (3) better detection limits for some elements than flame emission, and (4) ease of operation.

Disadvantages

The disadvantages of AAS are: (1) chemical interferences present for elements that form stable compounds; (2) ionization enhancement of the signal for elements easily ionized when operating in the absorption mode, especially Na and K; (3) matrix interferences caused by viscosity or specific gravity differences between sample and reference standard; (4) much less linearity than for inductively coupled plasma emission spectrometry; (5) elements analyzed one at a time; and (6) not capable for the determination for phosphorus (P) and boron (B).

INDUCTIVELY COUPLED PLASMA ATOMIC EMISSION SPECTROMETRY AND DIRECT CURRENT PLASMA EMISSION SPECTROMETRY

Principle of Operation and Use

This instrument measures the emission of radiant energy from excited atoms or ions. It uses the principle that when a neutral or partially ionized atom is excited by an energy source, valence electrons of the atom enter higher energy orbits around the atom's nucleus. These electrons return to ground-state energy orbits in the cooler regions of the excitation source. When this occurs, increments of electromagnetic radiation are given off with wavelengths characteristic of the atoms that are excited. Light emission instruments use grating monochromators or polychromators to separate these wavelengths so that the detector system can detect them. Intensity value for the emitted energy is directly proportional to the element concentration. The intensity value is collected and processed by a computer system. Details on the principle operation and use of the inductively coupled plasma atomic emission spectrometry (ICP-AES) have been described in the articles by Dalquist and Knoll (1978), Soltanpour et al. (1982), and Sharp (1991) and in the books by Thompson and Walsh (1983), Boumans (1984), Zarcinas (1984), Montaser and Golightly (1987), and Varma (1991).

The ICP-AES instrument is the principal instrument in most contemporary plant analysis laboratories (Watson and Isaac, 1990). Plasma source instruments are generally used to determine P, K, Ca, Mg, Zn, Fe, Cu, Mn, and B in plant tissue. Other elements such as heavy metals can also be determined by ICP-AES. Inductively coupled plasma atomic emission spectrometers that are either evacuated or purged with nitrogen gas can detect emission lines in the ultraviolet (UV) region of the spectra, being able to utilize the more sensitive P and B emission lines as well as the element sulfur (S). The sample is injected into the plasma through a nebulizer system, which separates the large droplets, allowing the small droplets to enter the plasma. Various kinds of nebulizers can be used, depending on the sample matrix and the detection limits required.

Excitation Sources

The term *DC plasma* refers to a hot gas in which a significant percentage of atoms have been ionized (Skogerboe et al., 1976). Inductively created electrical current is used to form and maintain the plasma. Argon (Ar) gas is used in most

ICP instruments; however, N gas has also been used for specific analytical purposes. Argon gas allows for an inert atmosphere surrounding the excited atoms and ions. Temperature of the Ar plasma ranges from 5,000 to 8,000°K (Fassel and Kniseley, 1974). This temperature allows complete vaporization of the sample, causing formation of free atoms and ions in the plasma.

Direct current plasma (DCP) instruments have also been used for plant analysis (Debolt, 1980). The excitation source in DCP instruments is the direct electrical current imparted to the Ar gas to form and maintain the plasma state. This is done with electrical current arcing across two or more electrodes in the gas flow.

Diffraction gratings are used to separate the wavelengths, photomultiplier tubes for converting light energy to electrical energy, and readout systems under computer control for digitizing the electrical signal. Different types of gratings can be used in spectrometers. In addition, there are usually two general types of spectrometers. Spectrometers that contain polychromators are used for simultaneous determinations of many elemental concentrations. Scanning monochromators are used for sequential determination of elements. In the simultaneous instruments, specific photomultiplier cubes must be present for each element of interest. In sequential instruments, one or two photomultiplier tubes are used and are not restricted to specific elements. Most modern plant-analysis laboratories use the simultaneous instruments due to their speed of analysis. However, the new sequential instruments allow more rapid analysis than earlier versions.

Advantages

The advantages of the ICP-AES are: (1) minimum chemical interferences, (2) four to six orders of magnitude in linearity of intensity versus concentration, (3) multielement capabilities, (4) rapid analysis, (5) accurate and precise analysis, and (6) detection limits equal to or better than AAS for many elements.

Direct current plasma instruments can essentially claim the same advantages as ICP-AES instruments, with the exception of reduced sensitivity of some elements by an order of magnitude less than for ICP-AES.

Disadvantages

Disadvantages of the ICP-AES are: (1) occurrence of spectral interferences, (2) use of argon gas which can be expensive, and (3) relatively expensive to purchase.

X-RAY FLUORESCENCE SPECTROMETRY

Principle of Operation and Use

The fundamental principles of X-ray fluorescence are explained by Kubota and Lazar (1971), Murdock and Murdock (1977), Dixon and Wear (1964), Skoog (1985), Jones (1991), Prange and Schwemke (1992), Günter et al. (1992), and Markert et al. (1994). The deceleration of high-energy electrons, or electronic transitions of inner orbital electrons of atoms, produces short wavelength electromagnetic radiation termed X-rays. X-ray fluorescence is used for the qualitative and quantitative analysis of elements with atomic numbers greater than oxygen (>8); and it has been used in plant analysis for all elements having atomic numbers greater than 11. The three basic types of X-ray fluorescence instruments are: (1) wavelength dispersive, (2) energy dispersive, and (3) non-dispersive. Energy-dispersive instruments are used for plant analysis.

Energy-dispersive instruments are capable of detecting more than 80 elements. X-ray fluorescence has been used to determine the concentrations P, K, Ca, Mg, Mn, Fe, Cu, Zn, Si, S, and Cl in plant tissue. The X-ray energies are too low for the determination of the concentrations of N and B in plant tissue. Approximately 100 plant tissue samples can be routinely analyzed per day for 10 or more elements (Knudsen et al., 1981).

Advantages

Results of X-ray fluorescence analysis of plant tissue compare favorably with those of AAS or ICP-AES (Iron et al., 1976; Knudsen et al., 1981; Günter et al., 1992). An important advantage is that the analysis is non-destructive and consequently rapid. Another advantage is that spectral line interferences are unlikely. The instrument is simple to operate. The closeness of the detector to the sample allows for the use of weak X-ray sources.

Disadvantages

The main disadvantage is the lack of standards made of a matrix the same or similar to plant tissue. In addition, samples should be ground to pass a 40-mesh sieve and a pellet must be made. Also, due to the low X-ray energies, B cannot be determined.

POTENTIOMETRIC INSTRUMENTS

Principle of Operation and Use

Potentiometric instruments involve the use of ion-selective electrodes and corresponding meters. Potentiometric methods have been reviewed by Carlson and Keeney (1971) and Street and Peterson (1982). A membrane of the ion electrode separates the external sample solution from the internal solution, which contains the ion of interest. Electrical current is carried through the membrane by a single ion species. The ion electrode is read against a reference electrode, which completes the electrical circuit. The membrane potential is logarithmically related to the ionic activity of the ion of interest. The membrane must have the properties of: (1) membrane material must have minimal solubility in analyte solutions; i.e., approaching zero; (2) membrane must exhibit small electrical conductivity, and (3) membrane or some ionic species contained in the membrane matrix must bind selectively with the analyte ion.

The nitrate electrode is the most common ion-selective electrode used for plant analysis. Nitrate electrode methods for measuring the concentration of nitrate-nitrogen in plant tissue have been reported by Paul and Carlson (1968), Baker and Smith (1969), Raveh (1973), and Sweetsur and Wilson (1975). The most common extracting solution for extracting nitrate-nitrogen from plant tissue is $0.025M$ aluminum sulfate $[Al(SO_4)_3]$. Chloride-specific electrodes have also been used to measure the concentration of chloride in plant tissue (Moody and Thomas, 1977; Krieg and Sung, 1977). Gas-sensing electrode specific for 1M has been used to measure N in Kjeldahl digests of plant tissue (Eastin, 1976).

Advantages

An advantage of using potentiometric instruments is the speed of determination. Also these instruments are simple to operate and relatively inexpensive.

Disadvantages

Numerous interferences can occur if the concentration of the interfering ionic species is substantially greater than the ion of interest in the sample. For example, chloride ions can interfere with the measurement of the nitrate ion concentration if the chloride concentration is high relative to that of the nitrate concentration. Malic and citric acids have been reported to interfere with the determination of the concentration of chloride when the chloride electrode is used (Watson and Isaac, 1990). Ions of interest must be extracted from the plant tissue before the selective electrodes can be used. Seldom do the standard samples match the unknown sample in ionic strength, thus requiring the use of a total ionic-strength adjustment buffer (TISAB) solution.

SULFUR ANALYZERS

Principle of Operation and Use

The advent of the sulfur analyzers for measuring the quantity of S in plant tissue has made the analysis for this element relatively simple. A certain quantity of sample (usually from 0.1 to 1.0 g) is weighed into a tared crucible and catalyst added. The crucible and sample is placed into a hot resistance furnace. Oxygen flows around the sample while it is heated to approximately 3,000°K. The S in the sample is converted to sulfur dioxide (SO_2). The gas stream carrying the SO_2 is scrubbed for halogen gases and water vapor and the quantity of SO_2 is measured with an infrared detector. The use of sulfur analyzers is discussed in detail in Chapter 12.

Advantages

Sulfur analyzers are simple and easy to use. The analysis is rapid, requiring approximately two minutes per sample. No chemicals other than the catalyst are used and no accelerators are needed. The instrument parameters are set through a computer system, sample weight is recorded by the computer, and the computer performs the necessary calculations to determine the S concentration in the plant sample.

Disadvantages

Furnace operates at high temperatures and oxygen gas is required. Vanadium pentoxide is usually used as the catalyst. Standards that are used to calibrate the instrument should be of an organic matrix similar to the plant tissue samples.

NITROGEN ANALYZERS

Principle of Operation and Use

In this chapter, nitrogen analyzers refer to those instruments that use a dry combustion process to analyze for nitrogen (N). The principle is based on the

Dumas method (Dumas, 1831). In nitrogen analyzers, the plant tissue sample is heated in a resistance furnace in a stream of oxygen (O_2). The N in the sample is converted to the oxides of N, which are carried through the instrument system in a stream of carrier [either carbon dioxide (CO_2) or helium (He)] gas. The carrier gas carries the oxidized N through warmed copper filings where the nitrogen oxide is reduced to N_2 gas. The determination of the quantity of N_2 is done by a thermal conductivity detector. Prior to the N determination, the carrier gas stream is scrubbed of halogen gases and moisture vapor is removed. The use of nitrogen analyzers for conducting an N determination in plant tissues is becoming more prevalent, replacing the traditional Kjeldahl procedure (see Chapter 9).

Advantages

The advantages and disadvantages of nitrogen analyzers is relative to the comparison with the traditional Kjeldahl determination. Nitrogen Analyzers are much less caustic to the environment, have fewer chemical wastes of which to dispose, and are cleaner and safer to operate than Kjeldahl digestion systems. Precision of the analysis compares favorably with Kjeldahl (Sweeney and Rexroad, 1987; Watson and Isaac, 1990; Simonne et al., 1994). Analytical accuracy is comparable to Kjeldahl for samples not containing nitrate (NO_3) and superior for samples containing more than 0.2% N. The N analysis is completed within 3 to 10 minutes and most instruments are controlled with a computer. The use of large sample trays allows for unattended N analysis.

Disadvantages

Disadvantages are the use of high temperature furnaces, oxygen gas, and changing scrubber and copper reduction columns frequently. Crucibles or stainless steel sample holders are required. The maintenance requirement is fairly high.

CONTINUOUS FLOW ANALYZERS: AIR-SEGMENTED INSTRUMENTS

Principle of Operation and Use

Various types of automated continuous flow analyzers have been used for plant analysis for at least 25 years (Steckel and Flannery, 1968, 1971; Basson et al., 1969; Isaac and Jones, 1970). These instruments consist of linking together separate modules to form a continuous single or multichemistry flow system.

The modules generally are the sampler, peristaltic pump, mixing manifold, colorimeter, and data recorder. These instruments may also include dialysis membranes, delay coils, and temperature baths. Computers are used to manage data and to make calculations. In the air-segmented instrument, air bubbles are added to the analytical stream so that portions of it are segmented. Air bubbles aid in mixing the reagent chemicals. The color reactions are developed in the mixing coils and the air bubble is physically or electronically removed before stream enters the colorimeter. The instrument has been used for the determination of the concentrations of P, K, Ca, Mg, Cl and SO_2 in wet digests of plant tissue (Watson and Isaac, 1990).

Advantages

Improved analytical precision is the main advantage over manual determinations since each sample has the same time for color development. In addition, little attention is needed once the instrument is started. The analysis is rapid, generally ranging from 20 to 40 samples per hour, depending on the element of interest.

Disadvantages

A disadvantage is that relatively new pump tubes must be used. Chemical interferences can occur, but usually are minimized by using complexing agents. A preparatory dissolution treatment of plant tissue is required.

CONTINUOUS FLOW ANALYZERS: NON-SEGMENTED STREAM INSTRUMENTS

Principle of Operation and Use

The operation of non-segmented stream instruments is similar to air-segmented stream instruments, essentially containing the same modules as the non-segmented stream instruments with the exception of the inclusion of an injection valve (Ranger, 1981; Rusicka and Hansen, 1988). The main features of the non-segmented stream instruments are: (1) controlled sample dispersion, (2) variable flow rates, (3) baseline resolution between each sample, (4) high sample throughput, and (5) absence of reaction stabilization time. The sample solution is injected directly into the carrier stream. Concentrations gradients (slugs) are formed in the carrier stream and are carried to the detector. Detectors can be ion-selective electrodes, AAS, ICP-AES, and colorimeters. The system relies on the very accurate and precise injection of the sample solution by the injector valve. The elements that can be determined are essentially the same as the air-segmented stream instrument; however, this will depend on the type of detector used.

Advantages

The analysis is very fast, 100 to 300 samples per hour. The start-up and shut-down times are minimal. The instrument can handle small sample solution sizes of 10 to 30 microliters. The instrument is precise, accurate, and simple to operate.

Disadvantages

The main disadvantage is that the injection valves must be durable and resistant to wear.

ION-EXCHANGE CHROMATOGRAPHY

This instrument is not commonly used for the routine analysis of plant tissue because of the importance of having a complete predigestion of the sample to remove all organic matter that may tend to plug the exchange column. However, it has been used successfully by some people for the determination of sulfate (SO_4), chloride (Cl), nitrate (NO_3), potassium (K), calcium (Ca), and magnesium (Mg) (Watson and Isaac, 1990).

NEAR INFRARED REFLECTANCE SPECTROMETRY

Principle of Operation and Use

The instrument uses monochromatic light directed at the plant tissue sample. Diffuse light is deflected from the sample and detected by lead sulfide detectors. Some instrument models use scanning monochromators, others use wavelength filters to isolate specific wavelengths of reflected radiation. The wavelength region is 1,100 to 2,500 nm for most scanning near infrared reflectance (NIR) instruments. The reflectance data are usually transformed to log_{10}/reflectance intensity and plotted as a function of wavelength to obtain reflectance spectra. The instrument is most often used to determine the quantity of protein and N in plant tissue. In order to calibrate the NIR instrument, it is necessary to determine the same parameters via wet chemistry on a large population of samples. Regression analysis is used to relate the wet chemistry values to reflectance intensity values at specific regions of the reflected spectra.

Advantages

The major advantage is that the analysis is non-destructive and is very rapid. Since no chemicals are needed, the analysis is safe. The instrument is simple to operate.

Disadvantages

The analysis is only as accurate as the wet chemistry results used to calibrate the instrument. In addition, a large population of sample must be used to obtain a good calibration. It is necessary to closely monitor the drift in the instrument since the recalibration is quite involved.

Instrumental Methods of Analysis

Given below is a compilation of some of the commonly used, past and present, analytical techniques for elemental and ion determination in prepared plant tissue extracts, digests, and ash solutions based on their suitability.

Element	Colorimetric	Emission Flame	Spark	ICP	X-ray	Atomic absorption	Specific-ion electrode
Boron (B)	Good	NA	Good	Ex	Poor	NA	NA
Calcium (Ca)	Good	Fair	Good	Ex	Poor	Ex	Poor
Copper (Cu)	Good	NA	Good	Ex	Ex	Ex	NA
Iron (Fe)	Fair	NA	Good	Ex	NA	Ex	NA
Magnesium (Mg)	Fair	Fair	Ex	Ex	Poor	Ex	NA
Manganese (Mn)	Good	NA	Ex	Ex	Poor	Ex	NA
Molybdenum (Mo)	Good	NA	Poor	Good	Good	Good[a]	NA
Phosphorus (P)	Ex	NA	Ex	Ex	Fair	NA	NA
Potassium (K)	Poor[b]	Ex	Ex	Ex	NA	Good	NA
Sodium (Na)	NA	Ex	Ex	Ex	NA	Good	Fair
Zinc (Zn)	Good	NA	Ex	Ex	Good	Ex	NA
Nitrate (NO_3)	Ex	NA	NA	NA	NA	NA	Good
Ammonium	Good	NA	NA	NA	NA	NA	Good
Chloride	Good	NA	NA	NA	NA	NA	Good
Fluoride	NA	NA	NA	NA	NA	NA	Good
Sulfate	Good[b]	NA	NA	NA	Good	NA	NA

Note: NA = not applicable; Ex = excellent (high sensitivity with minimal interference); Good = moderate sensitivity with some interference; Fair = reasonable sensitivity but with matrix effects; Poor = reasonable sensitivity with significant matrix effects.

[a] Turbidity.
[b] Flameless AA.

REFERENCES

Baker, A.S. and R. Smith. 1969. Extracting solution for potentiometric determination of nitrate in plant tissue. *J. Agric. Food Chem.* 17:1284–1287.

Baker, D.L. and N.H. Suhr. 1982. Atomic absorption and flame emission spectrometry, pp. 13–27. In: A.L. Page (Ed.), *Methods of Soil Analysis.* Part 2. Chemical and Microbiological Properties. Agronomy Monograph No. 9. American Society of Agronomy, Madison, WI.

Barnes, R.M. (Ed.). 1991. *Developments in Atomic Plasma Spectrochemical Analysis.* Heyden, London, England.

Basson, W.D., R.G. Boehmer, and D.A. Stanton. 1969. An automated procedure for the determination of boron in plant tissue. *Analyst* 94:1135–1141.

Boumans, P.W.J.M. (Ed.). 1984. *Inductively Coupled Plasma Emission Spectroscopy.* Part II: Applications and Fundamentals. John Wiley & Sons, New York.

Carlson, R.M. and D.R. Keeney. 1971. Specific ion electrodes: Techniques and uses in soil, plant, and water analysis, pp. 39–65. In: L.M. Walsh (Ed.), *Instrumental Methods for Analysis of Soils and Plant Tissue.* Soil Science Society of America, Madison, WI.

Dalquist, R.L. and J.W. Knoll. 1978. Inductively coupled plasma emission spectroscopy: Analysis of biological materials and soil for major, trace, and ultra-trace elements. *Appl. Spectrosc.* 32:1–30.

Debolt, D.C. 1980. Multielement emission spectroscopic analysis of plant tissue using DC argon plasma source. *J. Assoc. Off. Anal. Chem.* 63:802–805.

Dixon, J.B. and J.I. Wear. 1964. X-ray spectrographic analysis of zinc, manganese, iron, and copper in plant tissue. *Soil Sci. Soc. Am. Proc.* 28:744–746.

Dumas, J.B.A. 1831. Procedes de l'analyse organique. *Ann. Chim. Phys.* 47:198–205.

Eastin, E.F. 1976. Use of ammonia electrode for total nitrogen determination in plants. *Commun. Soil Sci. Plant Anal.* 7:477.

Fassel, V.A. and R.N. Kniseley. 1974. Inductively coupled plasmas. *Anal. Chem.* 46:1155A–1164A.

Günter, K. A. von Bohlen, G. Paprott, and R. Klockenhämper. 1992. Multielement analysis of biological reference materials by total-reflection X-ray fluorescence and inductively coupled plasma emission spectrometry in the semiquant mode. *Fresenius Z. Anal. Chem.* 342:444–448.

Helrich, K. (Ed.). 1995. *Official Methods of Analysis of the Association of Official Analytical Chemists.* Volume 1. Method 985.01. Association of Official Analytical Chemists, Arlington, VA.

Iron, R.D., E.A. Schenk, and R.D. Glauque. 1976. Energy-dispersive x-ray fluorescence spectroscopy and inductively coupled plasma emission spectrometry evaluated for multielement analysis in complex biological matrices. *Clin. Chem.* 22:2018–2024.

Isaac, R.A. and W.C. Johnson, Jr. 1975. Collaborative study of wet and dry ashing techniques for elemental analysis of plant tissue by atomic absorption spectrophotometry. *J. Assoc. Off. Anal. Chem.* 58:436–440.

Isaac, R.A. and W.C. Johnson, Jr. 1985. Elemental analysis of tissue by plasma emission spectroscopy: Collaborative study. *J. Assoc. Off. Anal. Chem.* 68:499.

Isaac, R.A. and J.B. Jones. 1970. Autoanalyzer systems for the analysis of soil and plant tissue extracts. *Adv. Auto. Anal.* 2:57–64.

Isaac, R.A. and J.D. Kerber. 1971. Atomic absorption and flame photometry: Techniques and uses in soil, plant, and water analysis, pp. 17–37. In: L.M. Walsh (Ed.), *Instrumental Methods for Analysis of Soils and Plant Tissue.* Soil Science Society of America, Madison, WI.

Jones, A.A. 1991. X-ray fluorescence spectrometry, pp. 287–324. In: K.A. Smith (Ed.), *Soil Analysis: Modern Instrumental Techniques.* Second edition. Marcel Dekker, New York.

Jones, Jr., J.B. and V.W. Case. 1990. Sampling, handling, and analyzing plant tissue samples, pp. 389–427. In: R. L. Westerman (Ed.), *Soil Testing and Plant Analysis.* Third edition. SSSA Book Series Number 3. Soil Science Society of America, Madison, WI.

Knudsen, D., R.B. Clar, J.L. Denning, and P.A. Pier. 1981. Plant analysis of trace elements by X-ray. *J. Plant Nutr.* 3:61–75.

Krieg, D.R. and D. Sung. 1977. Interferences in chloride determination using the specific ion electrode. *Commun. Soil Sci. Plant Anal.* 8:109–114.

Kubota, J. and V.A. Lazar. 1971. X-ray emission spectrograph: Techniques and uses for plant and soil studies, pp. 67–82. In: L.M. Walsh (Ed.), *Instrumental Methods for Analysis of Soils and Plant Tissue.* Soil Science Society of America, Madison, WI.

Market, B., U. Reus, and U. Herpin. 1994. The application of TXRF in instrumental multielement analysis of plants, demonstrated with species of moss. *Sci. Total Environ.* 152:213– 230.

Markert, B. 1995. Instrumental Multielement Analysis in Plant Materials—A Modern Method in Environmental Chemistry and Tropical Systems Research. Technologia Ambiental. MCT. CNpq CETEM. Rio de Janeiro, Brazil.

Metcalfe, E. 1987. *Atomic Absorption and Emission Spectrometry.* John Wiley & Sons, New York.

Montaser, A. and D.W. Golightly (Eds.). 1987. *Inductively Coupled Plasmas in Analytical Atomic Spectrometry.* VCH Publishers, New York.

Moody, G.J. and J.D.R. Thomas. 1977. The determination of chloride in vegetables, fruits and juices with ion-selective electrode. *J. Food Technol.* 12:193–197.

Murdock, A and O. Murdock. 1977. Analysis of plant material by X-ray fluorescence spectrometry. *X-ray Spectros.* 6:215–217.

Paul, J.L. and R.M. Carlson. 1968. Nitrate determination in plant extracts by the nitrate electrode. *J. Agric. Food Chem.* 16:766–768.

Prange, A. and H. Schwemke. 1992. Trace element analysis using total reflection X-ray fluorescence spectrometry. *Adv. X-ray Anal.* 35:899–923.

Rains, T.C. and O. Menis. 1976. An intercomparison of flame and nonflame systems in atomic absorption spectrometry, pp. 1045–1051. In: P.D. LaFluer (Ed.), *Accuracy in Trace Analysis: Samples, Sample Handling, Analysis.* Volume 2. Special Publication 422. U.S. Department of Commerce/NBS, Gaithersburg, MD.

Ranger, C.B. 1981. Flow injection analysis: Principles, techniques, application and design. *Anal. Chem.* 53:20A–27A.

Raveh, A. 1973. The adaptation of the nitrate-specific electrode for soil and plant analysis. *Soil Sci.* 116:388–389.

Rusicka, J. and E.H. Hanson. 1988. *Flow Injection Analysis.* Second edition. John Wiley & Sons, New York.

Sharp, B.L. 1991. Inductively coupled plasma spectrometry, pp. 63–109. IN: K.A. Smith (Ed.), *Soil Analysis: Modern Instrumental Techniques.* Second Edition. Marcel Dekker, New York.

Simonne, E.H., H.A. Mills, J.B. Jones, Jr., D.A. Smittle, and C.G. Hussey. 1994. Comparison of analytical methods for nitrogen analysis in plant tissues. *Commun. Soil Sci. Plant Anal.* 25:943–954.

Skogerboe, R.K., L.T. Urasa, and G.N. Colemen. 1976. Characterization of a DC plasma as an excitation source for multielement analysis. *Appl. Spectros.* 30:500–507.

Skoog, D.A. 1985. *Principles of Instrumental Analysis.* Third edition. Saunders College, New York.

Soltanpour, P.N., J.B. Jones, Jr., and S.M. Workman. 1982. Optical emission spectrometry, pp. 29–65. In: A.L. Page et al. (Ed.), *Methods of Soil Analysis.* Part 2. Second edition. Agronomy Monograph Number 9. American Society of Agronomy, Madison, WI.

Steckel, J.E. and R.L. Flannery. 1968. Automatic determination of phosphorus, potassium, calcium, and magnesium in wet digestion solutions of plant tissue. *Tech. Qual.* 1:19–20.

Steckel, J.E. and R.L. Flannery. 1971. Simultaneous determinations of phosphorus, potassium, calcium, and magnesium in wet digestion solutions of plant tissue by autoanalyzer, pp. 83–96. In: L.M. Walsh (Ed.), *Instrumental Methods for Analysis of Soils and Plant Tissue.* Soil Science Society of America, Madison, WI.

Street, J.J. and W.M. Peterson. 1982. Anodic stripping voltammetry and differential pulse polarography, pp. 133–148. In: A.L. Page (Ed.), *Methods of Soil Analysis.* Part 2. Chemical and Microbiological Properties. Agronomy Monograph No. 9. American Society of Agronomy, Madison, WI.

Sweeney, R.A. and P.R. Rexroad. 1987. Comparison of LECO FR-228 "Nitrogen determinator" with AOAC copper catalyst Kjeldahl method for crude protein. *J. Assoc. Off. Anal. Chem.* 70:1028–1030.

Sweetsur, A.W.M. and A.G. Wilson. 1975. An ion-selective electrode method for the determination of nitrate in grass and clover. *Analyst* 100:485–488.

Thompson, M. and J.N. Walsh. 1983. *A Handbook of Inductively Coupled Plasma Spectrometry*. Blackie, Glasgow, Scotland.

Tsalev, D.L. 1984. *Atomic Absorption Spectrometry in Occupational and Environmental Health Practice*. Volume II. CRC Press, Boca Raton, FL.

Ure, A.M. 1991. Atomic absorption and flame emission spectrometry, pp. 1–62. In: K.A. Smith (Ed.), *Soil Analysis: Modern Instrumental Techniques*. Second edition. Marcel Dekker, New York.

Varma, A. 1991. *CRC Handbook of Inductively Coupled Plasma Emission Spectrometry*. CRC Press, Boca Raton, FL.

Walsh, A. 1955. Application of atomic absorption spectra to chemical analysis. *Spectrochim. Acta* 7:108–117.

Walsh, L.M. (Ed.). 1971. *Instrumental Methods for Analysis of Soils and Plant Tissue*. Soil Science Society of America, Madison, WI.

Watson, M.E. and R.A. Isaac. 1990. Analytical instruments for soil and plant analysis, pp. 691–740. In: R. L. Westerman (Ed.), *Soil Testing and Plant Analysis*. Third. edition. SSSA Book Series Number 3. Soil Science Society of America, Madison, WI.

Zarcinas, B.A. 1984. *Analysis of Soil and Plant Material by Inductively Coupled Plasma-Optical Emission Spectrometry*. CSIRO. Division of Soils. Divisional Report No. 70. Commonwealth Scientific and Industrial Research Organization, Melbourne, Australia.

DETERMINATION OF POTASSIUM AND SODIUM BY FLAME EMISSION SPECTROPHOTOMETRY

19

Donald A. Horneck and Dean Hanson

INTRODUCTION

In the last 20 years, the use of a flame emission spectrophotometer as an analytical procedure for the determination of potassium (K) and sodium (Na) in a plant tissue digest has declined since atomic absorption and plasma emission spectrometers have become the instruments of choice for the determination of these and other elements. A major limitation for a flame emission spectrophotometer is that relatively few elements, primarily K, Na, and lithium (Li), are the only elements easily determined, compared to either an atomic absorption or plasma emission spectrophotometer, although flame emission instruments are relatively inexpensive, have a wide working concentration range (0 to 100 mg/L), have excellent sensitivity (0.0005 µg/L) and precision characteristics, and can use natural gas as the fuel for the flame. Most atomic absorption spectrophotometers can be also operated in the emission mode (see Chapter 20).

1-57444-124-8/98/$0.00+$.50
© 1998 by CRC Press LLC

PRINCIPLE OF OPERATION

Basically, a flame spectrophotometer consists of a sample introduction system, an excitation source, either an acetylene/air or natural gas/air flame burner and attached nebulizer, means of producing monochromatic light (interference filters, prism, or grating), and detector. Within the flame, an atom of an element is transformed into an ion by the loss of shell electrons. With the recapture of lost electrons, energy is given off in the form of light (photons) of characteristic wavelength, which can be used to identify the element, while the intensity of the light is directly proportional to the concentration of the element in the flame. The analyte containing the element(s) of interest must be in solution, which is then atomized into the flame. Further information on the principles of operation and use of both techniques can be found in either the articles by Isaac and Kerber (1971), Baker and Shur (1982), Watson and Isaac (1990), and Ure (1991) and the books by Mavrodineanu (1970) and Metcalfe (1987).

EQUIPMENT

1. Flame emission spectrophotometer, wavelength settings, K at 766.5 nm and Na at 589.0 nm.
2. Digestion facilities (see Chapters 5 through 8).

REAGENTS

1. **Potassium (K) Primary Standard** (1,000 mg/L): weigh 1.9067 g oven-dried and desiccated potassium chloride (KCl) and quantitatively transfer to a 1-L volumetric flask and bring to volume with deionized water.
2. **Sodium (Na) Primary Standard** (1,000 mg/L): weigh 2.5421 g desiccated sodium chloride (NaCl) into a 1-L volumetric flask. Dissolve with 8 mL deionized water and 8 mL concentrated HCl. Bring to volume with additional deionized water.
3. **Working Standards:** prepare as series of working standards from the primary standards within the normal operating range of the flame emission spectrophotometer, normally between 0 to 100 mg/L.

PROCEDURE

1. Digest plant tissue using one of the procedures outlined in Chapters 5 through 8.

2. Dilute the digest or ash to the volume that will put the elements to be determined within the detection range of the spectrophotometer.
3. Following the procedures given for the flame spectrophotometer employed, calibrate and then assay the prepared plant digests.

COMMENTS

1. Primary elemental standards for K, Na, and Li can be obtained from commercial sources.
2. Frequently, Li is added to the standards and analyte at a background concentration of approximately 2,000 mg/L to prevent ionization prior to aspiration into the flame as well as serving as an internal standard if the spectrometer is equipped for using an internal standard.
3. Calcium (Ca) and magnesium (Mg) can be determined by flame emission if the analyte does not contain sizable concentrations (>50 mg/L) of other elements, particularly the elements iron (Fe) and aluminum (Al), and the anions sulfate (SO_4) and phosphate (PO_4), unless these ions are removed from the analyte or a compensation solution is added to bring these ions into equal concentration among the standards and analytes.

REFERENCES

Baker, D.L. and N.H. Shur. 1982. Atomic absorption and flame emission spectrometry, pp. 13-27. In: A.L. Page (Ed.), *Methods of Soil Analysis*. Part 2. Chemical and Microbiological Properties. Agronomy Monograph No. 9. American Society of Agronomy, Madison, WI.

Isaac, R.A. and J.D. Kerber. 1971. Atomic absorption and flame photometry: Techniques and uses in soil, plant, and water analysis, pp. 17-37. In: L.M. Walsh (Ed.), *Instrumental Methods for Analysis of Soils and Plant Tissue*. Soil Science Society of America, Madison, WI.

Mavrodineanu, R. (Ed.). 1970. *Analytical Flame Spectroscopy: Selected Topics*. Springer-Verlag, New York.

Metcalfe, E. 1987. *Atomic Absorption and Emission Spectrometry*. John Wiley & Sons, New York.

Ure, A.M. 1991. Atomic absorption and flame emission spectrometry, pp. 1-62. In: K.A. Smith (Ed.), *Soil Analysis: Modern Instrumental Techniques*. Second edition. Marcel Dekker, New York.

Watson, M.E. and R.A. Isaac. 1990. Analytical instruments for soil and plant analysis, pp. 691-740. In: R.L. Westerman (Ed.), Soil Testing and Plant Analysis. Third edition. SSSA Book Series Number 3. Soil Science Society of America, Madison, WI.

ELEMENTAL DETERMINATION BY ATOMIC ABSORPTION SPECTROPHOTOMETRY

20

Edward A. Hanlon

INTRODUCTION

The elements aluminum (Al), calcium (Ca), copper (Cu), iron (Fe), magnesium (Mg), potassium (K), sodium (Na), and zinc (Zn) in a plant tissue digest brought into solution by one of several procedures for organic matter destruction (see Chapters 5–8) can be determined by atomic absorption spectrophotometry (AAS). The plant tissue liquid digest containing the elements to be determined is atomized into either an acetylene/air or acetylene/nitrous oxide gas mixture at a temperature between 2,000· to 2,900°C. The burner design and adjustment of the fuel/oxidant mixture provide conditions in which the elements to be determined are converted to non-excited, non-ionized, ground-state atoms.

PRINCIPLE OF OPERATION

Atomic Absorption Spectrophotometry

An atomic absorption spectrophotometer consists of a sample introduction system, an excitation source (hollow cation lamp), nebulizer and flame burner, chopper, and detector. An atom of an element is capable of absorbing light energy characteristic of that element. Radiation (photons) is generated from a hollow cathode lamp whose cathode is made of the element for determination. When these photons pass through the flame containing atoms of the element, the photons are absorbed. The degree of absorption is proportional to the concentration of the element in the flame (the flame also serves as a means of supporting the atoms in the light path). The measured difference between the light intensity passing around the flame and that passing through the flame defines absorption and can be used to determine the concentration of the element in the atomized solution. The analyte containing the element(s) of interest must be a solution that can be atomized into the flame.

Flame Emission Spectrophotometry

A flame emission spectrophotometer consists of a sample introduction system, an excitation source, either an acetylene/air or natural gas/air flame burner, means of producing monochromatic light (interference filters, prism, or grating), and detector. Most AAS instruments can be operated in either the atomic absorption or flame emission modes. Within the flame, an atom of an element is transformed into an ion by the loss of shell electrons. Energy is also given off in the form of light (photons) of characteristic wavelength that can be used to identify the element, while the intensity of the light is directly proportional to the concentration of the element in the atomized solution. The analyte solution containing the element(s) of interest must be a solution that can be atomized into the flame. The flame emission spectrophotometer technique is discussed in Chapter 19 and a brief overview of both methods is discussed in Chapter 18.

Measurement, whether by atomic absorption or flame emission spectrophotometry, is usually done with a movable monochromator. The light path is controlled by specifically designed slits and photodetectors that are sensitive to the wavelength(s) in question (Hanlon, 1992a,b). Further information on the principles of operation and use of both techniques can be found in either the articles by Isaac and Kerber (1971), Baker and Shur (1982), Watson and Isaac (1990), and Ure (1991), and the books by Tsalva (1984) and Metcalfe (1987).

EQUIPMENT

1. Atomic absorption/flame emission spectrophotometer.
2. Digestion facilities: perchloric-acid hood and wet chemistry digestion equip-

ment. Alternately, a muffle furnace and suitable containers for use in the furnace (see Chapters 5–8). Other digestion methods can be used (Isaac and Johnson, 1975; White and Douthit, 1985).

REAGENTS

The following text includes instructions for preparing primary standards for use with atomic absorption/flame emission spectrometry. However, certified primary elemental standards can be purchased from commercial sources for preparing working standards. In addition to reducing the technical workload, these commercially available standards can be a part of a sound quality-assurance program within the laboratory (see Chapter 25).

1. Hydrochloric acid (12M HCl), concentrated reagent grade.
2. Nitric acid (16N HNO$_3$), concentrated reagent grade.
3. Deionized water.
4. **Lanthanum (La) Solution** (1,000 mg/L): prepared from either lanthanum oxide (La$_2$O$_3$) or from lanthanum chloride (LaCl$_3$·H$_2$O) containing 0.1M HCl. The La$_2$O$_3$ must be brought into solution using HCl, but is much cheaper than the more readily soluble hydrated chloride source. Using La$_2$O$_3$, prepare a slurry by adding a small volume of deionized water to 1.1727 g La$_2$O$_3$ in a 1-L volumetric flask. Slowly add 8 mL concentrated HCl and stir. Dilute to final volume with additional deionized water. Starting with LaCl$_3$·H$_2$O, dissolve 2.6738 g LaCl$_3$·H$_2$O in deionized water. Slowly add 8 mL of concentrated HCl and bring to volume with additional deionized water.
5. **Potassium (K) Standard** (1,000 mg/L): weigh 1.9067 g oven-dried and desiccated potassium chloride (KCl), quantitatively transfer to a 1-L volumetric flask, and bring to volume with deionized water.
6. **Calcium (Ca) Standard** (1,000 mg/L): weigh 2.4973 g oven-dried and desiccated calcium carbonate (CaCO$_3$) into a 1-L volumetric flask. Slowly add (dropwise) approximately 8 mL concentrated HCl. Bring to volume with deionized water.
7. **Magnesium (Mg) Standard** (1,000 mg/L): weigh 1.000 g Mg metal ribbon into a 1-L volumetric flask, dissolve with 8 mL deionized water and 8 mL concentrated HCl. Bring to volume with additional deionized water.
8. **Manganese (Mn) Standard** (1,000 mg/L): weigh 1.000 g Mn metal into a 1-L volumetric flask. Dissolve with a minimum of dual parts deionized

water and HNO_3. Add 8 mL HCl and bring to volume with additional deionized water.

9. **Iron (Fe) Standard** (1,000 mg/L): weigh 1.000 g Fe wire into a 1-L volumetric flask. Dissolve with approximately 8 mL deionized water and 8 mL concentrated HCl. Bring to volume with deionized water.

10. **Copper (Cu) Standard** (1,000 mg/L): weigh 1.000 g Cu metal into a 1-L volumetric flask. Dissolve with 8 mL deionized water and 8 mL concentrated HCl. Bring to volume with additional deionized water.

11. **Zinc (Zn) Standard** (1,000 mg/L): weigh 1.000 g Zn metal ribbon into a 1-L volumetric flask. Dissolve with 8 mL deionized water and 8 mL concentrated HCl. Bring to volume with additional deionized water.

12. **Aluminum (Al) Standard** (1,000 mg/L): weigh 1.000 g Al metal ribbon into a 1-L volumetric flask. Dissolve with 8 mL deionized water and 8 mL concentrated HCl. Bring to volume with additional deionized water.

13. **Sodium (Na) Standard** (1,000 mg/L): weigh 2.5421 g desiccated sodium chloride (NaCl) into a 1-L volumetric flask. Dissolve with 8 mL deionized water and 8 mL concentrated HCl. Bring to volume with additional deionized water.

PROCEDURE

After performing either a dry ash or wet acid digestion on a known dry weight (usually about 1 g) of tissue, the obtained ash or digest is wetted with a small amount of deionized water and then brought into solution using 2 mL concentrated HCl. The final dilution with deionized water should be based on the predicted concentration of the element to be determined, ensuring that the final concentration is neither at or below the method detection limits nor above the normal operation range. For determination of the elements K, Ca, Mg, Al, and Na, a 100-mL final volume should provide concentrations sufficiently above the detection limit for a 1.0 g sample of most plant materials. For determination of the elements Mn, Fe, Cu, and Zn, final volumes between 10 to 50 mL are required.

For example, Isaac and Kerber (1971) gave typical elemental concentration ranges for plant tissue, with the range for Cu being 1 to 25 mg/kg dry plant tissue. Although the published detection limit by AAS is 0.002 µg Cu/mL (Perkin-Elmer Corp., 1976), such a low detection limit should never be assumed, but determined for the instrument to be used. A daily working detection limit of 0.05 µg Cu/mL is more realistic for high volume laboratory operations.

Using the latter detection limit and selecting a final volume of 50 mL, results reported in the tissue would have a lower limit of 2.5 mg/Cu/kg dry tissue (0.05 × 50 mL/1 g tissue). Should lower levels be expected, the final dilution volume should be reduced. The amount of statistical uncertainty surrounding a measurement increases sharply when the reading approaches a concentration within ten times the method detection limit (Taylor, 1989).

After bringing to final volume, the solution should be mixed by inversion of the volumetric flask several times.

Potassium may be determined either by AAS or flame emission spectrophotometry in this solution, although flame emission spectrophotometry has been reported to be somewhat more sensitive (Perkin-Elmer Corp., 1976). Serial dilutions should be made until the K concentration reading is within the standardized range of the instrument using 0.1 to $0.3M$ HCl as the diluent. The actual linear range of K is between 0 and 10 mg K/L. However, commercial instrumentation can be programmed with three to five serial dilutions of the K standard to extend the upper limit of the working range to between 50 to 100 mg K/L.

For determinations of Ca and Mg by AAS, an aliquot of the prepared analyte solution should be diluted with the La standard using 9 mL 1,000 mg La/L for every 1 mL of analyte solution. The linear working range for Ca is from 0 to 10 mg Ca/L, while that of Mg is 0 to 0.5 mg/L. These ranges can often be extended, depending on the capabilities of the instrument, as was discussed for K. Typical working ranges are approximately 0 to 50 mg Ca and Mg/L.

The published upper linear ranges are Zn = 1 mg/L, Mn = 3 mg/L, and Cu and Fe = 5 mg/L (Perkin-Elmer Corp., 1976). With curve-fitting ability, which is available on most atomic absorption spectrophotometers, a useful upper working limit of 10 mg/L can be achieved with good accuracy and precision (Hanlon et al., 1994).

The preferred method of analysis is by atomic absorption for Fe and Zn. Either atomic absorption or flame emission can be used for Cu and Mn (Isaac and Kerber, 1971). Since these elements are often determined as a group, all four are usually best determined by AAS (Isaac and Jones, 1972).

Aluminum has a nominal working range of 10 to 150 mg/L. However, Al must be analyzed in a nitrous oxide/acetylene (N_2O/C_2H_2) flame.

Sodium has a working range of 0.3 to 3 mg/L in an air/C_2H_2 flame and is normally best determined in the emission mode.

The wavelength setting, concentration range, and sensitivity for the elements are as follows:

| | | μg/mL | |
Element	Wavelength (nm)	Concentration range in solution	Detection limit
Aluminum (Al)	309.2	10–150	0.100
Calcium (Ca)	422.6	1–10	0.002
Copper (Cu)	324.7	2–20	0.005
Iron (Fe)	248.3	2–20	0.005
Magnesium (Mg)	285.2	0.1–0.2	0.0003
Potassium (K)	766.4	1–10	0.005
Sodium (Na)	589.0	0.3–3	0.002
Zinc (Zn)	213.8	0.2–3	0.002

COMMENTS

1. The various analytical approaches used to prepare and analyze plant tissue for elemental assay have been reviewed by Jones and Case (1990) and Jones (1991).
2. Potassium, Ca, and Mg appear to have lower variability when atomic absorption methods are used compared to flame emission spectrophotometry.
3. The selection of a digestion procedure does not necessarily mean that the choice of a diluent is clear. For example, several methods of bringing the resulting dry ash into solution have been described (Isaac and Kerber, 1971; Isaac and Jones, 1972; Munter and Grande, 1981; Campbell and Plank, 1992; Hanlon et al., 1994).
4. In general, a small volume of a strong acid, such as HCl or HNO_3, is used, regardless of the digestion technique, to enhance elemental solubility. The resulting solution is then diluted, most often with deionized water, producing a final acidic solution between 0.1 and 0.3M, depending on the acid and final dilution. Acidification also provides biological control resulting from the lowered pH. The procedure described above uses HCl.

INTERFERENCES

Potassium

Partial ionization of K occurs in an air/acetylene flame. Addition of 1,000 mg of other alkali salts has been reported to decrease such ionization (Perkin-Elmer Corp., 1976).

Calcium

Silicon, P, Al, and sulfate depress Ca absorption. The addition of La in excess greatly reduces these anionic effects as well as suppressing Ca ionization interferences. The literature reports use of La in concentrations from 0.1% to as high as 5%. However, excellent results have been obtained using the procedure described above for plant material with a substantial savings of La salts.

Magnesium

The elements Si and Al depress Mg absorption. The addition of La in excess greatly reduces these effects. There are no reported interferences for the determination of Mn, Cu, and Zn by AAS.

The presence of HNO_3 and Ni or Si may depress the sensitivity of Fe. Use of a very hot, lean-burning, air/acetylene flame appears to overcome these interferences (Isaac and Kerber, 1971; Perkin-Elmer Corp., 1976).

Both Al and Na can be affected by ionization interferences, which can be minimized by addition of 1,000 mg KCl/L (Isaac and Kerber, 1971).

REFERENCES

Baker, D.L. and N.H. Shur. 1982. Atomic absorption and flame emission spectrometry, pp. 13–27. In: A.L. Page (Ed.), *Methods of Soil Analysis*. Part 2. Chemical and Microbiological Properties. Agronomy Monograph No. 9. American Society of Agronomy, Madison, WI.

Campbell, C.E. and C.O. Plank. 1992. Sample preparation, pp. 1–1.1. In: C.O. Plank (Ed.), *Plant Analysis Reference Procedures for the Southern Region of the United States*. Southern Cooperative Series Bulletin 368. University of Georgia, Athens.

Hanlon, E.A. 1992a. Determination of potassium, calcium, and magnesium in plants by atomic absorption techniques, pp. 33–36. In: C.O. Plank (Ed.), *Plant Analysis Reference Procedures for the Southern Region of the United States*. Southern Cooperative Series Bulletin 368. University of Georgia, Athens.

Hanlon, E.A. 1992b. Determination of total manganese, iron, copper, and zinc in plants by atomic absorption techniques, pp. 49–51. In: C.O. Plank (Ed.), *Plant Analysis Reference Procedures for the Southern Region of the United States*. Southern Cooperative Series Bulletin 368. University of Georgia, Athens.

Hanlon, E.A., J.S. Gonzales, and J.M. Bartos. 1994. IFAS Extension Soil Testing Laboratory Chemical Procedures and Training Manual. Circular No. 812 (revised). Florida Cooperative Extension Service, Institute Food Agricultural Science, University of Florida, Gainesville.

Isaac, R.A. and W.C. Johnson, Jr. 1975. Collaborative study of wet and dry ashing techniques for elemental analysis of plant tissue by atomic absorption spectrophotometry. *J. Assoc. Off. Anal. Chem.* 58:436–440.

Isaac, R.A. and J.B. Jones, Jr. 1972. Effects of various dry ashing temperatures on the determination of 13 nutrient elements in five plant tissues. *Commun. Soil Sci. Plant Anal.* 3:261–269.

Isaac, R.A. and J.D. Kerber. 1971. Atomic absorption and flame photometry: Techniques and uses in soil, plant, and water analysis, pp. 17–37. In: L.M. Walsh (Ed.), *Instrumental Methods for Analysis of Soils and Plant Tissue*. Soil Science Society of America, Madison, WI.

Jones, Jr., J.B. 1991. Plant tissue analysis in micronutrients, pp. 477–521. In: J.J. Mortvedt (Ed.), *Micronutrients in Agriculture*. SSSA Book Series No. 4. Soil Science Society of America, Madison, WI.

Jones, Jr., J.B. and V.W. Case. 1990. Sampling, handling, and analyzing plant tissue samples, pp. 389–427. In: R.L. Westerman (Ed.), *Soil Testing and Plant Analysis*. Third edition. SSSA Book Series No. 3. Soil Science Society of America, Madison, WI.

Metcalfe, E. 1987. *Atomic Absorption and Emission Spectrometry*. John Wiley & Sons, New York.

Munter, R.C. and R.A. Grande. 1981. Plant tissue and soil extract analysis by ICP-atomic emission spectrometry, pp. 653–672. In: R.M. Barnes (Ed.), *Developments in Atomic Plasma Spectrochemical Analysis*. Heyden and Sons, London, England.

Perkin-Elmer Corp. 1976. *Analytical Methods for Atomic Absorption Spectrophotometry*. Perkin-Elmer Corp., Norwalk, CT.

Taylor, J.K. 1989. *Quality Assurance of Chemical Measurements*. Lewis Publishers, Chelsea, MI.

Tsalva, D.L. 1984. *Atomic Absorption Spectrometry in Occupational and Environmental Health Practice*. CRC Press, Boca Raton, FL.

Ure, A.M. 1991. Atomic absorption and flame emission spectrometry, pp. 1–62. In: K.A. Smith (Ed.), *Soil Analysis: Modern Instrumental Techniques*. Second edition. Marcel Dekker, New York.

Watson, M.E. and R.A. Isaac. 1990. Analytical instruments for soil and plant analysis, pp. 691–740. In: R.L. Westerman (Ed.), *Soil Testing and Plant Analysis*. Third edition. SSSA Book Series Number 3. Soil Science Society of America, Madison, WI.

White, Jr., R.T. and G.E. Douthih. 1985. Use of microwave oven and nitric acid/hydrogen peroxide digestion to prepare botanical materials for elemental analysis by inductively coupled argon plasma emission spectroscopy. *J. Assoc. Off. Anal. Chem.* 68:766–769.

ELEMENTAL DETERMINATION BY INDUCTIVELY COUPLED PLASMA ATOMIC EMISSION SPECTROMETRY

21

Robert A. Isaac and William C. Johnson, Jr.

INTRODUCTION

The determination of the elements boron (B), calcium (Ca), copper (Cu), iron (Fe), magnesium (Mg), manganese (Mn), phosphorus (P), potassium (K), and zinc (Zn) in a plant tissue digest or ash solution (see Chapters 5–8) by inductively coupled plasma atomic emission spectrometry (ICP-AES) is described. The principles of operation and application of the technique have been discussed by Jones (1977), Dalquist and Knoll (1978), Munter and Grande (1981), Soltanpour et al. (1982), Thompson and Walsh (1983), Zarcinas (1984), Boumans (1984), Isaac and Johnson (1985), Montaser and Golightly (1987), Varma (1991), and Sharp (1991). Munter et al. (1984) have presented the quality assurance requirements for the ICP-AES technique.

PRINCIPLE OF OPERATION

A plasma is simply a very hot gas (internal temperature up to 10,000 K) in which a significant portion of the atoms or molecules is ionized. The ICP source produces a stream of high-energy ionized gas by inductively coupling an inert gas (commonly argon) with a high-frequency field. When a fine mist of liquid sample is introduced into the center of the plasma, it desolvates, dissociates, atomizes, and excites the elements in the sample. When the generated ions return to their atomic (ground) state, light (photons) is emitted, the wavelength identifying the element and the intensity of light elemental concentration. The basic principle of an ICP for analytical application has been described by Fassel and Kniseley (1974).

The high temperature of the ICP source in an ICP-AES has several advantages over other emission-type analyzers:

- Excitation of refractory and rare-earth elements.
- Five to seven orders of linearity; wide dynamic concentration range.
- Very low detection limits (exceptional sensitivity), mid parts per billion (ppb).
- No to limited chemical interferences.
- Minimal interelement effects.
- Excellent accuracy and precision.
- High performance-to-cost ratio.

Therefore, an ICP-AES has the ideal qualities for the easy assay of plant tissue digests for all but a few of the elements [namely nitrogen (N)] essential for plants, plus those elements, at trace or ultra-trace levels found in plants although not essential (Dalquist and Knoll, 1978). Sulfur (S) can also be determined if the spectrometer is evacuated or purged with nitrogen (N_2) gas, making usable wavelength lines in the ultraviolet portion of the spectrum.

Spectrometers are of two primary designs: polychromators, in which each determinable element has its own exit slit and detector installed in the spectrometer; and the sequential spectrometer, in which either the grating or detector is moved in a programmed sequence from one wavelength position to another during the integration procedure. For high-speed replicated assay work, the polychromator is the preferred design; for assay work that does not fit a fixed elemental requirement, the sequential design offers flexibility in element selection.

Elemental wavelength selection is based on emission line strength and lack of potential spectral interference. Commonly selected wavelengths for the elements given in this chapter are:

Element	Wavelength (nm)	Element	Wavelength (nm)
Boron (B)	249.7	Manganese (Mn)	257.6
Calcium (Ca)	422.6	Phosphorus (P)	214.9×2[a]

Element	Wavelength (nm)	Element	Wavelength (nm)
Copper (Cu)	327.4	Potassium (K)	766.4
Iron (Fe)	259.9	Zinc (Zn)	213.8
Magnesium (Mg)	279.5		

[a] Second order position.

Wavelength tables and line characteristics may be found in the book by Varma (1991).

EQUIPMENT

1. Inductively coupled plasma emission spectrometer.
2. 30-mL glazed porcelain crucible.
3. 50-mL volumetric flasks.
4. Muffle furnace.
5. Hot plate.
6. Source of argon (Ar) gas-liquid is preferred.

REAGENTS

1. **Nitric Acid (1+1):** add 250 mL conc. nitric acid (HNO_3) to a 500-mL volumetric flask containing 250 mL deionized water and mix well.
2. **Hydrochloric Acid (1+1):** add 250 mL conc. hydrochloric acid (HCl) to a 500-mL volumetric flask containing 250 mL deionized water and mix well.
3. **Primary Element Standards:** these standards can be obtained from commercial sources or made from reagents as follows:
   - **Boron (B) Standard** (1,000 mg/L): weigh 5.715 g boric acid (H_3BO_3) into a 1-L volumetric flask and bring to volume with deionized water.
   - **Calcium (Ca) Standard** (10,000 mg/L): weigh 24.973 g oven-dried and desiccated calcium carbonate ($CaCO_3$) into a 1-L volumetric flask. Slowly add approximately 80 mL concentrated HCl. Bring to volume with deionized water.
   - **Copper (Cu) Standard** (1,000 mg/L): weigh 1.000 g Cu metal into a 1-L volumetric flask. Dissolve with 8 mL deionized water and 8 mL concentrated HCl. Bring to volume with additional deionized water.
   - **Iron (Fe) Standard** (1,000 mg/L): weigh 1.000 g Fe wire into a 1-L volumetric flask. Dissolve with approximately 8 mL deionized water and 8 mL concentrated HCl. Bring to volume with deionized water.

- **Magnesium (Mg) Standard** (10,000 mg/L): weigh 10.000 g Mg metal ribbon into a 1-L volumetric flask and dissolve with 8 mL deionized water and 80 mL concentrated HCl. Bring to volume with additional deionized water.
- **Manganese (Mn) Standard** (1,000 mg/L): weigh 1.000 g Mn metal into a 1-L volumetric flask. Dissolve with a minimum of dual parts deionized water and HNO_3. Add 8 mL HCl and bring to volume with additional deionized water.
- **Phosphorus (P) Standard** (10,000 mg/L): weigh 42.640 g oven-dried and desiccated dibasic ammonium phosphate $(NH_4)_2HPO_4$ and quantitatively transfer to a 1-L volumetric flask and bring to volume with deionized water.
- **Potassium (K) Standard** (10,000 mg/L): weigh 19.067 g oven-dried and desiccated potassium chloride (KCl) and quantitatively transfer to a 1-L volumetric flask and bring to volume with deionized water.
- **Zinc (Zn) Standard** (1,000 mg/L): weigh 1.000 g Zn metal ribbon into a 1-L volumetric flask. Dissolve with 8 mL deionized water and 8 mL concentrated HCl. Bring to volume with additional deionized water.

4. **Calibration Standards:** prepare these standards from the primary standards as follows:
    - Standard No. 1 (concentration in mg/L): Cu 5; Fe 5; K 1,000; Mg 200; Mn 2; P 20.
    - Standard No. 2 (concentration in mg/L): B 2; Zn 5.
    - Standard No. 3 (concentration in mg/L): Ca 100.

SAMPLE PREPARATION PROCEDURE

1. Weigh 1.00 ± 0.05 g plant tissue into a 30-mL glazed porcelain crucible.
2. Place the crucible into a cool muffle furnace and muffle at 500°C for 2 hours.
3. Remove the crucible from muffle furnace, let cool, and add 3.0 mL HNO_3 (1+1).
4. Heat sample on a hot plate at 100 to 120°C until dry.
5. Place the crucible back in to the muffle furnace and muffle at 500°C oven for 1 additional hour.
6. Remove crucible from the muffle furnace, let cool, and add 10 mL HCl (1+1).
7. Transfer sample to a 50-mL volumetric flask. Dilute to volume with deionized water and mix well.

ANALYSIS PROCEDURE

Analyze the obtained plant tissue ash solution in accordance with AOAC Method
985.01 (Helrich, 1995) and/or based on recommendations provided by the instrument manufacturer.

COMMENTS

1. It is desirable to use one supplier to produce a set of calibration standards
   and another supplier to produce a set of instrument calibration verification
   standards (Munter et al., 1984). Reference Plant Materials should be used
   to verify the accuracy of the elemental determinations made (see Chapters
   25, 26, and Appendix 1).
2. The elemental concentration range in the calibration standards should bracket
   the expected concentration to be found in the unknowns.
3. The selection of the elements to be included in a calibration standard and
   the selection of the blank for each element are important since the setting
   of the *zero* (no element present) will determine the minimum concentration
   detected and set the intercept of the calibration curve.
4. Although calibration line linearity is a characteristic of this technique, it
   needs to be verified for those elements (i.e., Ca, K, and Mg) that range in
   concentration over several orders of magnitude.
5. Depending on the spectrometer and characteristics of the plasma, the use of
   an internal standard may improve the precision of the assay for some elements (Jones, 1977).
6. In general, the focus should be on precision since most elements in a plant
   tissue digest or dissolved ash solution are considerably above the detection
   limit of the ICP-AES technique, so sensitivity is not a major problem, while
   repeatability over time can be.

References

Baker, D.L. and N.H. Shur. 1982. Atomic absorption and flame emission spectrometry,
   pp. 13–27. In: A.L. Page (Ed.), *Methods of Soil Analysis.* Part 2. Chemical and
   Microbiological Properties. Agronomy Monograph No. 9. American Society of
   Agronomy, Madison, WI.
Boumans, P.W.J.M. 1984. *Inductively Coupled Plasma Emission Spectroscopy.* Part II:
   Applications and Fundamentals. John Wiley & Sons, New York.

Dalquist, R.L. and J.W. Knoll. 1978. Inductively coupled plasma atomic emission spectroscopy: Analysis of biological materials and soil for major, trace, and ultra-trace elements. *Appl. Spectrosc.* 32:1–30.

Fassel, V.A. and R.N. Kniseley. 1974. Inductively coupled plasmas. *Anal. Chem.* 46:1155A–1164A.

Helrich, K. (Ed.). 1995. *Official Methods of Analysis of the Association of Official Analytical Chemists.* Volume 1. Method 985.01. Association of Official Analytical Chemists, Arlington, VA.

Isaac, R.A. and W.C. Johnson, Jr. 1985. Elemental analysis of tissue by plasma emission spectroscopy: Collaborative study. *J. Assoc. Off. Anal. Chem.* 68:499.

Jones, Jr., J.B. 1977. Elemental analysis of soil extracts and plant tissue ash by plasma emission spectroscopy. *Commun. Soil Sci. Plant Anal.* 8:345–365.

Metcalfe, E. 1987. *Atomic Absorption and Emission Spectrometry.* John Wiley & Sons, New York.

Montaser, A. and D.W. Golightly (Eds.). 1987. *Inductively Coupled Plasmas in Analytical Atomic Spectrometry.* VCH Publishers, New York.

Munter, R.C. and R.A. Grande. 1981. Plant tissue and soil extract analysis by ICP-atomic emission spectrometry, pp. 653–672. In: R.M. Barnes (Ed.), *Developments in Atomic Plasma Spectrochemical Analysis.* Heyden and Sons, London, England.

Munter, R.C., T.L. Halverson, and R.D. Anderson. 1984. Quality assurance of plant tissue analysis by ICP-AES. *Commun. Soil Sci. Plant Anal.* 15:1285–1322.

Sharp, B.L. 1991. Inductively coupled plasma spectrometry, pp. 63–109. In: K.A. Smith (Ed.), *Soil Analysis: Modern Instrumental Techniques.* Second edition. Marcel Dekker, New York.

Soltanpour, P.N., J.B. Jones, Jr., and S.M. Workman. 1982. Optical emission spectrometry, pp. 29–65. In: A.L. Page et al. (Eds.), *Methods of Soil Analysis.* Part 2. Second edition. Agronomy Monograph Number 9. American Society of Agronomy, Madison, WI.

Thompson, M. and J.N. Walsh. 1983. A Handbook of Inductively Coupled Plasma Spectrometry. Blackie, Glasgow, Scotland.

Varma, A. 1991. *CRC Handbook of Inductively Coupled Plasma Emission Spectrometry.* CRC Press, Boca Raton, FL.

Watson, M.E. and R.A. Isaac. 1990. Analytical instruments for soil and plant analysis, pp. 691–740. In: R.L. Westerman (Ed.), *Soil Testing and Plant Analysis.* Third edition. SSSA Book Series Number 3. Soil Science Society of America, Madison, WI.

Zarcinas, B.A. *1984. Analysis of Soil and Plant Material by Inductively Coupled Plasma-Optical Emission Spectrometry.* CSIRO. Division of Soils. Divisional Report No. 70. Commonwealth Scientific and Industrial Research Organization, Melbourne, Australia.

DETERMINATION OF BORON, MOLYBDENUM, AND SELENIUM IN PLANT TISSUE

22

Umesh C. Gupta

INTRODUCTION

The two micronutrients boron (B) and molybdenum (Mo), and the element selenium (Se) are more difficult to determine in plant tissues than some of the other micronutrients and trace elements found in plants. Very low concentrations (<0.1 mg/kg) of Mo and Se in plants, which exist in most regions of the world, are principal causes for this difficulty (Pais and Jones, 1996). Boron is present in higher quantities (>10 mg/kg) in typical plant tissues than Mo or Se, elements difficult to determine by inductively coupled plasma atomic emission spectrometry (ICP-AES) because of their inadequate sensitivity compared to other elements but determinable by graphite-furnace atomic absorption spectrometry (GF-AAS) (Watson and Isaac, 1990). Boron is easily determined by ICP-AES (see Chapter 21).

This chapter describes the analytical techniques most commonly used for the determination of these three elements.

1-57444-124-8/98/$0.00+$.50
© 1998 by CRC Press LLC

BORON

Introduction

There are a number of techniques available for determining B in plant tissues. They include titrimetric, spectrometric, and colorimetric methods. The latter makes use of reagents, such as curcumin, turmeric, quinalizarin, or azomethine-H (Wolf, 1974; Johnson and Fixen, 1990; Offiah and Axley, 1993) for the color development. Today, the commonly used procedures include determination by ICP-AES (Watson and Isaac, 1990; Sharp, 1991) and the azomethine-H colorimetric method (Wolf, 1974; Gupta, 1979; Gupta and MacLeod, 1982).

Inductively Coupled Plasma Atomic Emission Spectrometry Method

Equipment

1. Silica crucibles.
2. Muffle furnace.
3. Inductively coupled plasma atomic emission spectrophotometer.
4. Centrifuge.

Reagents

1. Deionized Type 1 water.
2. Hydrochloric acid (2N HCl).
3. Standard Boron Solution
a. **Stock Solution** (1,000 mg B/L): prepare by weighing 5.715 g boric acid (H_3BO_3) into a 1-L volumetric flask and bringing to volume with deionized water.
b. **Working Standards:** prepare a series of working standards, 0.5 to 6 mg B/L, by appropriate dilutions of the 1,000 mg B/L stock standard.

Procedure

1. Weigh 1.0 g dried ground (2-mm sieve) plant sample into a silica crucible.
2. Place crucibles in a cold muffle furnace and ash at 500°C for 4 hours.
3. Remove the crucible from the muffle furnace, let cool, and then dissolve the ash in 5 mL 20% HCl.
4. Dilute solution to 50 mL with deionized water.
5. Centrifuge the solution in tubes at 3,000 cpm for 10 minutes or filter.

Analysis

1. The clear supernatant solution is analyzed for its B content by ICP-AES at wavelength 2497 nm for a vacuum or purged spectrometer, or at 2497 ¥ 2 nm for an air-path spectrometer.
2. Follow the manufacturer's recommendations for operating and calibrating the spectrometer.

Comments

The ICP-AES method is sufficiently sensitive to accurately detect B at concentrations normally found in plant tissue from deficiency to toxicity levels with excellent precision. Additional information on the ICP-AES analysis techniques may be found in Chapters 18 and 21.

Azomethine-H Colorimetric Method

Equipment

1. Silica crucibles.
2. Muffle furnace.
3. AutoAnalyzer or UV-VIS spectrophotometer.
4. Whatman No. 541 or 40 filter paper.
5. Volumetric flasks, 1-L and 100-mL, and pipettes.

Reagents

1. Deionized Type 1 water
2. Hydrochloric acid (2N IICl).
3. **Azomethine-H** (0.5% w/v): dissolve 0.5 g azomethine-H in 10 mL deionized water containing 1.0 g L-ascorbic acid by gentle heating at 30°C and make the volume up to 100 mL with deionized water.
4. **EDTA Reagent:** 0.25M [containing 4 g sodium hydroxide (NaOH)/L and 1 mL Brij-35/L].
5. **Buffer Reagent:** dissolve 250 g ammonium acetate ($NH_4C_2H_3O_2$) in 500 mL double deionized water, adjust to pH 5.8 by adding approximately 50 mL glacial acetic acid with constant stirring and add 0.5 mL Brij-35 solution and mix.
6. **Wash Solution** (0.4N HCl): dilute 33.3 mL conc. HCl to 1 L with deionized water.
7. **Standard Boron Solution**

a. **Stock Solution** (1,000 mg B/L): prepare by weighing 5.715 g boric acid (H_3BO_3) into a 1-L volumetric flask and bringing to volume with deionized water.

b. **Working Standards:** prepare a series of working standards, 0.5 to 6 mg B/L, by appropriate dilutions of the 1,000 mg B/L stock standard.

Procedure

1. Weigh 1.25 g dried ground (2-mm mesh sieve) plant material into a silica crucible.
2. Place crucibles in a cool furnace and ash at 500°C for 4 hours.
3. Remove the crucible from the muffle furnace, let cool, and then add 2.5 mL $2N$ HCl.
4. After 15 minutes, add 10 mL deionized Type 1 water to give a total volume of 12.5 mL.
5. Filter the solution using a Whatman No. 540 or 40 filter paper into a plastic vial.

Analysis

Analyze a portion of aliquot by the azomethine-H colorimetric method using an AutoAnalyzer. A simple spectrophotometer can also be used for the colorimetric determination by measuring the color absorbency at 430 nm. A detailed description of the reagents used and method of analyses are described above.

Comments

The use of azomethine-H is an improvement over that of carmine, quinalizarin, and curcumin, since the procedure involving this reagent does not require the use of a concentrated acid. The developed colored complex is stable for up to 4 hours.

MOLYBDENUM

Introduction

Molybdenum concentration in most plant materials is frequently less than 1 mg/kg; therefore, to detect Mo in trace quantities, the analytical procedure used must be very sensitive.

Molybdenum can be analyzed by GF-AAS (Curtis and Grusovin, 1985), by atomic absorption spectrometry (Khan et al., 1979), and by various colorimetric methods (Perrin, 1946; Bingley, 1961; Gupta and MacKay, 1965). Many of these methods are more effective for determining Mo in samples with high Mo values (>1 mg/kg) and plant extracts in some cases may require concentration. Since the dithiol colorimetric method as modified by Gupta and MacKay (1965) is simple, accurate, and inexpensive, and the Mo-dithiol colored complex is stable, that method will be described in detail, as well as the GF-AAS method, which is suitable for the determination of low levels of Mo in plant tissues.

Molybdenum Dithiol Colorimetric Method

Equipment

1. Silica crucibles.
2. Muffle furnace.
3. Hot plate.
4. Separatory funnels, 125-mL.
5. Centrifuge.
6. UV-VS spectrophotometer.
7. Burette for delivering concentrated sulfuric acid (H_2SO_4).
8. Gooch crucible fitted with a fiberglass filter disc.

Reagents

1. Deionized Type 1 water.
2. Sulfuric acid (H_2SO_4), concentrated.
3. Hydrogen peroxide (H_2O_2), 30%.
4. Ferrous ammonium sulfate [$FeSO_4 \cdot (NH_4)_2SO_4$], 9.1%.
5. Potassium iodide (KI), 50% w/v.
6. Ascorbic acid, 5% w/v.
7. Tartaric acid, 50% w/v.
8. Thiourea, 10% w/v.
9. Amyl acetate (B.P. 136 to 142°C).
10. Sodium hydroxide (NaOH), analytical grade.
11. **Toluene-3, 4-Zinc Dithiol Derivative:** prepare this reagent by adding 0.2 g zinc dithiol to 100 mL 1% NaOH, which has been warmed to 55°C and stir vigorously on a slightly warm hot plate with magnetic stirring for 10 minutes. While stirring, continuously add 1.7 mL thioglycolic acid. After 2 to 4 minutes of stirring, vacuum filter the solution through a Gooch crucible using fiberglass filter paper.

12. **Molybdenum Standard Solutions**
    a. **Stock Solution** (100 mg Mo/L): weigh 0.1508 g dried MoO_3 into a 150-mL beaker and add 3 mL $1N$ NaOH. Make the solution slightly acidic by adding 3.8 mL $1N$ HCl and make the volume up to 1,000 mL with deionized water.
    b. **Working Standard** (1 mg Mo/L): dilute 10 mL stock solution to 1,000 mL with deionized water. For samples containing high Mo concentrations, use a 10-mg Mo/L working standard.
    All reagents for Mo analyses should be prepared fresh.

Procedure

1. Weigh 2.0 g dried, ground (2-mm sieve) plant material into a silica crucible.
2. Place crucibles in a cold furnace and ash at 550°C for 4 hours.
3. Remove the crucible from the muffle furnace, let cool, and then add 4 mL concentrated H_2SO_4 to the crucible.
4. When the reaction with acid ceases, add 1.5 mL 30% H_2O_2 and swirl.
5. Place on a hot plate for 30 minutes and reheat with additional 1.5 mL H_2O_2 if the digest remains black, indicating incomplete destruction of organic matter.
6. When cool, transfer contents to a 50-mL volumetric flask and make to volume with deionized water.
7. Filter the contents using a Whatman No. 40 (11-cm) filter paper and analyze the total filtrate for Mo.

Analysis

1. Transfer all the filtered aliquot into a 125-mL separatory funnel, add 0.25 mL 9.1% $FeSO_4 \cdot (NH_4)_2SO_4$, 0.25 mL 50% KI, mix, and let stand for 15 minutes.
2. Add 0.25 mL 5% ascorbic acid and shake until the color disappears.
3. Add 0.25 mL 50% tartaric acid, shake, add 2 mL 10% thiourea, and mix thoroughly.
4. Add 4 mL 0.2% toluene-3,4-zinc dithiol derivative solution, shake for 20 minutes, and allow the contents to stand for 30 minutes.
5. Add 10 mL amyl acetate, shake vigorously for 2 minutes and allow to stand for 1 hour for complete separation.
6. Draw off and discard the aqueous phase.
7. Drain off the organic phase in a centrifuge tube and centrifuge for 15 minutes at 2,000 rpm.

8. Measure the color on a spectrophotometer at 680 nm using a red filter and sample blank as reference. From a prepared calibration curve, determine the amount of Mo in the digest.

Comments

The dithiol colorimetric method is suitable for determining Mo in plant tissues and other materials over a wide range of Mo concentrations. The method is sensitive for determining Mo at low concentrations (<0.1 mg Mo/kg), which are more common, than at high to toxic Mo concentrations. The method is inexpensive and can be used in most laboratories without difficulty.

Graphite Furnace Atomic Absorption Spectrophotometry Method

Equipment

1. Borosilicate glass tubes.
2. Aluminum block digester.
3. Centrifuge.
4. Varian GTA-95 graphite tube atomizer with a Varian model 875 atomic absorption spectrophotometer equipped with a deuterium arc lamp.
5. Argon (Ar) and nitrogen (N_2) gas.
6. Pyrolytic graphite tubes with partitioned walls.

Reagents

1. Deionized Type 1 water.
2. Nitric acid ($15M$ HNO_3).
3. Perchloric acid ($11M$ $HClO_4$).
4. Sulfuric acid $18M$ H_2SO_4).
5. Hydrochloric acid ($6.2M$ HCl).
6. Ammonium thiocyanate [$(NH_4)SCN$], 16% w/v.
7. Ammonia hydroxide ($17M$ NH_4OH).
8. **Stannous Chloride** (30% w/v): dissolve 30 g stannous chloride ($SnCl_2$) in 20 mL $6.2M$ HCl and boil until solution is clear. After cooling, dilute to 100 mL with distilled water. One or two granules of metallic tin (Sn) are added to this solution to prevent degradation.
9. **Standard Iron Solution:** dissolve 0.7022 g ferrous ammonium sulfate [$FeSO_4 \cdot (NH_4)_2SO_4$] in deionized water and dilute to 1 L with deionized water containing 1% v/v H_2SO_4.

10. **Mixed Reagent A:** mix 6.2M HCl and 30% w/v SnCl$_2$ in the ratio 5:1, respectively. Add 20 mL diisobutyl ketone (DIBK) and shake for 60 seconds. After settling, the reagent is separated from DIBK.

11. **Mixed Reagent B:** mix 16% w/v ammonium thiocyanate (NH$_4$)SCN and standard iron solution in the ratio 4:1, respectively. Add 20 mL DIBK to the mixture and shake for 60 seconds. After settling, separate the reagent from the DIBK.

12. **Standard Molybdenum Solution:** BDH standard solution containing 100 mg Mo/L was used as a stock solution and diluted to 10 mg Mo/mL and to 1 mg Mo/mL with deionized water as required.

Procedure

1. Weigh 1.0 g dried ground (0.5-mm mesh sieve) plant material into a boro-silicate glass tube.
2. Add 7 mL 15M HNO$_3$, 2 mL 11M HClO$_4$, and 1 mL 18M H$_2$SO$_4$ to the tubes.
3. Allow to stand overnight at 25°C and transfer to an aluminum block digester for complete digestion at 300 to 310°C.
4. Boil the mixture until fuming ceases and volume is reduced to about 1 mL.
5. After cooling, add 5 mL deionized water and boil to dissolve precipitated salts to destroy nitroso-sulfuric compounds.
6. Solutions are cooled and neutralized with 17M ammonia solution using methyl orange indicator.
7. Add 6.2M HCl dropwise till the solution turns just pink. Now add 6 mL mixed Reagent A and dilute to 32 mL and mix.
8. Now add 5 mL mixed Reagent B mix and add 5 mL DIBK and shake the solution vigorously for 40 seconds.
9. The solvent layer is then separated and centrifuged for 10 minutes before being transferred to sample cups.

Analysis

A sample volume of 10 µL is used for analyses by GF-AAS. Measurements are made in the peak height mode at a wavelength of 313.3 nm using a spectral bandwidth of 0.5 nm. Both N$_2$ and argon are used as the sheath gases. The calibration of the instrument is checked after every five samples to compensate for changes in the response due to tube deterioration.

Comments

This method improves the accuracy and precision for determining low Mo levels, the GF-AAS procedure is applicable to a range of different plant matrices. The instrument response for Mo concentrations ranging between 0 and 1 mg Mo/kg has been found to be linear.

SELENIUM

Introduction

Selenium (Se) deficiencies are more common than are toxicities (Kabata-Pendias and Pendias, 1995; Pais and Jones, 1996). Therefore, it is highly pertinent that the technique used is able to analyze for microquantities of Se. There are a number of techniques available for determining Se, e.g., fluorometry (Levesque and Vendette, 1971; Inhat, 1974), gas chromatography (Shimoishi, 1974; McCarthy et al., 1981), spectrophotometric analysis (Olson, 1973), and hydride generation and AAS (Clinton, 1977; Cox and Bibb, 1981; Brumbaugh and Walther, 1989; Beach, 1992). The selenium hydride (H_2Se) generation technique has been most commonly used over the last 15 years because it is accurate, rapid, and can detect plant Se in very low quantities without difficulty. This method incorporates the generation of H_2Se with sodium borohydride and conversion of the H_2Se to atomic Se with an electric heated absorption tube, which minimizes matrix problems and interferences associated with flame methods. Therefore, the hydride generation electrothermal atomization AAS technique will be described here in detail.

Equipment

1. 250-mL digestion tubes.
2. Digestion block (Tecator Digestion System 20 or equivalent equipment).
3. Perchloric acid fumehood.
4. Atomic absorption spectrophotometer equipped with hydride generation electrothermal unit.
5. Hollow cathode lamp for Se.

Reagents

1. Deionized Type 1 water.
2. Nitric acid ($15M$ HNO_3).

3. Hydrochloric acid (12M HCl).
4. Perchloric acid (11M HClO$_4$).
5. Hydrochloric acid (6M HCl).
6. **Selenium Standard Solutions**
a. **Stock Solution** (100 mg Se/L): weigh 0.1633 dried selenous acid (H$_2$SeO$_3$) in 200 mL deionized water, dilute to 1,000 mL, and store in a polyethylene bottle.
b. **Working Standard** (0.5 mg Se/L): pipette 5 mL stock solution into a 1,000-mL volumetric flask and dilute to volume with 6N HCl. Use 2 to 32 mL portions of this solution containing 1 to 16 mg Se for preparing a standard curve.

Procedure

1. Weigh 1.0 g dried ground (2-mm sieve) plant material into a digestion tube.
2. Add 10 mL concentrated HNO$_3$, adding in small portions swirling gently.
3. Now add 10 mL HClO$_4$ (70%) in small portions with gentle swirling and allow the tubes to sit in a perchloric acid fumehood overnight.
4. Add 5 mL concentrated HNO$_3$ down the sides of the tubes and place them in a cold digestion block and gradually raise temperature to 150°C in nine steps over a 4-hour period.
5. Cool the digest and add 10 mL 6M HCl to reduce Se^{6+} to Se^{4+}.
6. Reposition the tubes on the block and raise the temperature to 150°C for 15 minutes.
7. Remove the digestion tubes from the block, cool, and transfer the contents into a 50-mL flask.
8. Dilute the volume to 25 mL with 6M HCl and mix thoroughly.

Analysis

The final digest is analyzed for Se by AAS using a continuous flow hydride generator as described by Rothery (1984) at a wavelength of 1960.3 nm. Before reading the unknown plant digests for Se, standard Se solutions of known concentration are run to establish an absorbency curve. This results in recording the actual Se concentration in the plant extract directly.

Comments

This extract is stable for at least 14 days. During the 2-week period, the digested extracts should be kept refrigerated. The extract should be kept for 36 hours before taking Se measurements. This technique allows accurate determination of Se in concentrations as low as 0.02 mg Se/kg.

ACKNOWLEDGMENT

The author gratefully acknowledges the technical assistance of B.P. Stevenson in the development and/or modification of the procedures described. Contribution No. 842. Research Centre, Agriculture and Agri-Food Canada, Charlottetown, PEI, Canada.

REFERENCES

Beach, L.M. 1992. Determination of As, Sb, and Se in Difficult Environmental Samples by Hydride Generation. Publication No. AA-105. Varian, Fernando, CA.

Bingley, J.B. 1961. Molybdenum in plants and animals. Determination of molybdenum in biological materials with dithiol—Control of copper interference. J. Agric. Food Chem. 11:130–131.

Brumbaugh, W.G. and M.J. Walther. 1989. Determination of arsenic and selenium in whole fish by continuous flow hydride generation atomic absorption spectrophotometer. J. Assoc. Off. Anal. Chem. 72:484A86.

Clinton, O.E. 1977. Determination of selenium in blood and plant material by hydride generation and atomic absorption spectroscopy. Analyst 102:187–192.

Cox, D.H. and A.E. Bibb. 1981. Hydrogen selenide evolution-electrothermal atomic absorption method for determining nanogram levels of total selenium. J. Assoc. Off. Anal. Chem. 64:265–269.

Curtis, P.R. and J. Grusovin. 1985. Determination of molybdenum in plant tissue by graphite furnace atomic absorption spectrophotometry (GFAAS). Commun. Soil Sci. Plant Anal. 16:1279–1291.

Gupta, U.C. 1979. Some factors affecting the determination of hot-water-soluble boron from Podzol soils using azomethine-H. Can. J. Soil Sci. 59:241–247.

Gupta, U.C. and D.C. MacKay. 1965. Determination of Mo in plant materials using 4-methyl-1,2-dimercaptobenzene (dithiol). Soil Sci. 99: 414–415.

Gupta, U.C. and J.A. MacLeod. 1982. Effect of Sea Crop 16 and Ergostim on crop yield and plant composition. Can. J. Soil Sci. 62:527–532.

Inhat, M. 1974. Fluorometric determination of selenium in foods. J. Assoc. Off. Anal. Chem. 57:368–372.

Johnson, G.V. and P.E. Fixen. 1990. Testing soils for sulfur, boron, molybdenum, and chlorine, pp. 265–273. In: R.L. Westerman (Ed.), Soil Testing and Plant Analysis. SSSA Book Series 3. Soil Science Society of America, Madison, WI.

Kabata-Pendias, S.A. and H. Pendias. 1995. Trace Elements in Soils and Plants. Second edition. CRC Press, Boca Raton, FL.

Khan, S.U., R.O. Cloutier, and M. Hidiroglou. 1979. Atomic absorption spectroscopic determination of molybdenum in plant tissue and blood plasma J. Assoc. Off. Anal. Chem. 62:1062–1064.

Levesque, M. and E.D. Vendette. 1971. Selenium determination in soil and plant materials. Can. J. Soil Sci. 51:85–93.

McCarthy, T.P., B. Brodie, I.A. Milner, and R.F. Bevill. 1981. Improved method for selenium determination in biological samples by gas chromatography. J. Chromatography 225:9–16.

Offiah, O.O. and J.H. Axley. 1993. Soil testing of boron on acid soils, pp. 105–123. In: U.C. Gupta (Ed.), Boron and Its Role in Crop Production. CRC Press, Boca Raton, FL.

Olson, O.E. 1973. Simplified spectrophotometric analysis of plants for selenium. J. Assoc. Off. Anal. Chem. 56:1073–1077.

Pais, I. and J. B. Jones, Jr. 1996. Handbook of Trace Elements in the Environment. St. Lucie Press, Delray Beach, FL.

Perrin, D.D. 1946. Determination of molybdenum in soils. New Zealand J. Sci. Tech. 28A:183–187.

Rothery, E. 1984. VGA-76 Vapor Generation Accessory Operation Manual. Varian Techtron Pty. Limited, Mulgrave, Victoria, Australia.

Sharp, B.L. 1991. Inductively coupled plasma emission spectrometry, pp. 63–109. In: K.A. Smith (Ed.), Soil Analysis: Modern Instrumental Techniques. Second edition. Marcel Dekker, New York.

Shimoishi, Y. 1974. The gas chromatographic determination of selenium in plant material with 4-nitro-O-phenylenediamine. Bull. Chem. Soc. Japan 47:997–1002.

Watson, M.E. and R.A. Isaac. 1990. Analytical instruments for soil and plant analysis, pp. 691–740. In: R.L. Westerman (Ed.), Soil Testing and Plant Analysis. SSSA Book Series 3. Soil Science Society of America, Madison, WI.

Wolf, B. 1974. Improvements in the azomethine-H method for the determination of boron. Commun. Soil Sci. Plant Anal. 5:39–44.

DETERMINATION OF ARSENIC AND MERCURY IN PLANT TISSUE

23

Yoong K. Soon

ARSENIC

Introduction

Arsenic (As) commonly occurs in terrestrial plants within the concentration range of 0.02 to 7 mg As/kg (Bowen, 1979; Kabata-Pendias and Pendias, 1995; Pais and Jones, 1996). It has recently been proposed that As is required by animals, and possibly humans, in ultra-trace amounts (Nielsen, 1984; Pais and Jones, 1996), but as yet no such proposal has been made for higher plants.

Two main methods have been widely used for destruction of organic matter in biological samples in preparation for As analysis: (1) nitric (HNO₃)-perchloric (HClO₄) acid digestion (Jacobs et al., 1970) and (2) dry ashing with an alkaline flux (de Oliveira et al., 1983). Microwave oven-based wet digestion (Abu-Samra et al., 1975), and a HNO_3-H_2SO_4-V_2O_3 digestion (Uthe et al., 1974) have also been proposed. A wet alkaline oxidation method was recently proposed by Zhu and Tabatabai (1995). These recent developments need further corroboration as they appear to have the advantages of convenience, safety, and rapidity. A nitric-perchloric acid digestion adapted from Zasoski and Burau (1977) is described below, as it is a proven sample dissolution technique for As as well as for selenium (Se) by a slight modification at the end of the digestion.

The routine methods of As analysis involve the generation of arsine (AsH_3) and its measurement by atomic absorption spectrometry (AAS) (Thompson and Thoresby, 1977) or by colorimetry using either complexation with silver diethyldithiocarbamate (Merry and Zarcinas, 1982) or molybdenum blue color development (Small and McCants, 1961). The preferred method of measurement is AAS since it is more sensitive, simpler, and suffers from less interference than colorimetry. Graphite furnace AAS (GF-AAS) is a possibility but may suffer from matrix effects. The basic design of a hydride generation system (for generating AsH_3) with subsequent AAS measurement may be broken into three components: (1) generation of the hydride vapor, (2) collection and transfer of the hydride to the atomizer, and (3) measurement by AAS in a flame or electrically heated silica tube.

Equipment

1. An electrically heated aluminum block digester.
2. 100-mL digestion tubes (50-mL Folin-Wu tubes, 25 mm wide and 200 mm high may also be used).
3. Analytical balance, accurate to 1.0 mg.
4. Repipette dispensers.
5. Atomic absorption spectrophotometer fitted with an As hollow cathode lamp or electrodeless discharge lamp.
6. A continuous flow hydride generation system.

Reagents

1. Nitric acid (70% HNO_3), reagent grade or better.
2. Perchloric acid (70% $HClO_4$), reagent grade.
3. Hydrochloric acid (37% HCl), reagent grade or better.
4. Sodium tetrahydridoborate ($NaBH_4$), 98% purity pellet.
5. Sodium hydroxide (NaOH), 99% purity pellets.
6. **Standard Solution** (1,000 mg As/L): dissolve 0.660 g As_2O_3 in 100 mL $0.1M$ NaOH and diluting to 500 mL with $0.1M$ HCl. Prepare working standards as required by serial dilution to a range of 0 to 40 ng/mL in approximately the same matrix as sample solutions. Commercially available 1,000 mg As/L standard solution can be used.
7. 1% m/v $NaBH_4$ in 0.5% m/v NaOH, prepared daily.
8. Pre-reductant: $1M$ potassium iodide (KI) in 10% m/v ascorbic acid.
9. High purity nitrogen (N_2) gas.

Digestion Procedure (adapted from Zasoski and Burau, 1971)

1. Weigh 0.500 g dried, ground (<0.5-mm) plant tissue on a tared piece of cigarette paper. Fold the paper into a ball and drop into digestion tube containing two glass beads or anti-bumping granules (non-selenized).
2. Add 5 mL HNO_3 and leave overnight
3. Following the predigestion at room temperature, place tubes and contents in aluminum block digester heated to 90° (±10°)C. Heat for 1 hour or until evolution of copious fumes of nitrogen dioxide (NO_2) has subsided.
4. Remove from aluminum block digester and cool.
5. Add 2 mL $HClO_4$ and return to aluminum block digester.
6. Digest at 180°C until material has cleared and only wisps of white fumes are visible in the digestion tube. This usually requires 2 to 3 hours of heating. If contents of tube char, cool and add an additional 0.5 mL $HClO_4$ and continue heating to a clear solution. The solution may have a yellowish tint, which usually disappears on cooling and dilution.
7. Add 15 mL HCl and dilute with water almost to the 100-mL graduation mark. Add 1 mL of the KI pre-reductant [to reduce As(V) to As(III)]. Make up to volume and mix by several inversions of the tube and contents. Save overnight and analyze the solution the following day.

Analysis

Hydride generation-atomic absorption spectrometry (HG-AAS) of As can be performed on a batch or continuous flow basis (Brumbaugh and Walther, 1989). The latter is recommended since it is faster, simpler to operate, and more reliable than the batch technique. When a continuous flow system is used, a continuous integrable signal is produced rather than a transient peak. The acidified sample and the reductant ($NaBH_4$) solutions are taken in suitable proportions by a peristaltic pump, mixed in a reaction coil, and the gaseous hydride separated and swept by argon (Ar) or N_2 gas into a heated silica tube aligned in the optical path of the spectrophotometer.

The silica tube is heated either electrically or by a lean air/acetylene flame to 800 to 950°C. Sampling may be automated by an autosampler. The gas or electrically heated silica tube, hydride generation and delivery unit, and autosampler may be purchased from manufacturers such as Perkin-Elmer or Varian.

A suitable set of parameters for operating the Varian VGA 76 hydride generator in a continuous mode for As analysis is:

1. Sample/standard flow rate: 7 mL/min.
2. Acid (10M HCl) flow rate: 1 mL/min.
3. NaBH$_4$ reductant flow rate: 1 mL/min.
4. Nitrogen gas flow rate: 45 mL/min (at 250 kPa back pressure).

The wavelength used for As analysis is 193.7 nm and working standard solutions used for plant tissues digested according to the above procedure may be in the range of 0 to 50 ng As/mL.

Selenium (Se) analysis can be conveniently included in the above digestion by a slight modification. The digestion procedure is carried through to the beginning of Step 7. After adding 15 mL HCl, the contents of the digestion tubes are heated at 90°C for 20 minutes to pre-reduce Se(VI) to Se(IV). The oxidation state of As(V) is not affected by this treatment. Make up to volume (100 mL) and remove a 10-mL aliquot for subsequent Se analysis by HG-AAS with a transfer pipette. Add 1 mL of the KI pre-reductant to the solution remaining in the digestion tube, mix, and leave overnight to reduce As(V) to As(III).

Continuous flow hydride generation has been successfully combined with inductively coupled plasma atomic emission spectrometry (ICP-AES) for the simultaneous determination of As, Se, and antimony (Sb) (Nygaard and Lowry, 1982; Pretorius et al., 1992). The reader is referred to those papers and de Oliveira et al. (1983) and citations therein for details and variations in the technique. The detection limit for As is 1 to 2 ng/mL, nearly similar to HG-AAS. Incorporation of a condensation trap between the hydride generator and ICP spectrometer improved detection to sub-nanogram levels (Hahn et al., 1982). Use of ultrasonic nebulizers in ICP spectrometers has been reported to improve detection limits for direct sample aspiration technique by an order of magnitude, compared to an ICP-AES equipped with conventional nebulizers (Olson et al., 1977), i.e., detection limits are more or less comparable to HG-AAS.

Comments

1. The use of HClO$_4$ involves potential explosion hazards. The use of a perchloric acid fumehood is mandatory and pretreatment of sample with HNO$_3$ is essential. The reader who is not familiar with the use of HClO$_4$ is referred to reports by Smith (1953) and the Analytical Methods Committee (1959).
2. Because the HG-AAS technique is extremely sensitive, contamination is a highly potential source of error. Condition all new glassware by soaking in hot 1:1 HNO$_3$. Rinse all washed glassware and glass apparatus with 5% v/v HNO$_3$ followed by a final deionized water rinse and dry before use.
3. Residual HNO$_3$ from the digestion procedure can result in interference in

the determination of As. If this is suspected, pretreatment of the sample solution with urea solution is recommended (Beach, 1992). This is not normally a problem when the prescribed digestion method is followed.

4. The HG-AAS system has to be conditioned before making measurements by alternating the highest standard and blank until stable results are obtained.

5. Background correction is not generally required but should be tried initially for samples of uncertain matrix to ensure that the absorption signal is specific for As.

6. When samples are expected to contain less than 1 mg As/kg, Folin-Wu tubes are recommended for the digestion. In this case, 7.5 mL HCl is added at Step 7 of the digestion before diluting to 50 mL volume.

7. Reagent blanks should be run with all batches of samples.

MERCURY

Introduction

Mercury (Hg) is normally present in plant tissues in the concentration range of 30 to 700 µg Hg/kg (NRCC, 1979; Adriano, 1986; Kabata-Pendias and Pendias, 1995; Pais and Jones, 1996). In general, although the availability of soil Hg to plants is low, there is a tendency for absorbed Hg to accumulate in plant roots. Adriano's (1986) review indicates that Hg may accumulate in other plant parts under some circumstances and its transfer in the food chain is of some concern.

Most routine methods for the determination of total Hg in biological samples involve a two-step procedure: (1) the conversion of all bound Hg in the sample to Hg(II) by wet oxidation, and (2) the reduction of Hg(II) to Hg° vapor for analysis. A successful method for a variety of biological samples is the HNO_3-H_2SO_4-V_2O_5 digestion procedure of Malaiyandi and Barrette (1970) as modified by Deitz et al. (1973), and subsequently, by Knechtel and Fraser (1979). Because digestion with mineral acids alone may result in incomplete recovery of Hg, an oxidizing substance such as permanganate is often added to prevent losses of Hg (Uthe et al., 1970; van Delft and Vos, 1988). To further reduce losses of Hg, digestions should be done in closed vessels (van Delft and Vos, 1988) or long digestion tubes (Knechtel and Fraser, 1979). Microwave digestions of fish (Barrett et al., 1978), and soil and peat samples (van Delft and Vos, 1988) followed by cold vapor atomic absorption gave Hg values comparable to those determined by other methods such as neutron activation analysis. Further research into adapting the microwave oven digestion for Hg and other trace elements in plant samples is desirable since this digestion procedure has been shown to be suitable for the analysis of most plant nutrients, i.e., having multielemental capability.

The HNO_3-H_2SO_4-V_2O_5 digestion procedure is described because it is simple, rapid (30 to 35 minutes), reliable, and has undergone considerable testing and development. An aliquot of the digestate is then reduced in a hydride generation unit to liberate Hg vapor, which is measured in an unheated cell by atomic absorption. This vapor generation technique provides sensitivities approximately four orders of magnitude better than direct-aspiration flame AAS.

Equipment

1. Analytical balance, accurate to 1.0 mg or better.
2. 40-tube aluminum block digester.
3. 100-mL digestion tubes or 50-mL Folin-Wu tubes.
4. Repipette dispensers.
5. Atomic absorption spectrophotometer with continuous flow vapor generation accessories, including flow-through absorption cell and mercury hollow cathode lamp.

Reagents

1. Nitric acid (70% HNO_3), reagent grade or better.
2. Sulfuric acid (96% H_2SO_4), reagent grade or better.
3. Vanadium pentoxide (V_2O_5), reagent grade.
4. Sodium tetrahydridoborate ($NaBH_4$), reagent grade.
5. Nitrogen (N_2) gas, high purity.
6. **Mercury Standard Solution** (1,000 mg Hg/L): dissolve 0.3384 g mercuric chloride ($HgCl_2$) in 100 mL 12.5% (v/v) HNO_3 and dilute to 250 mL. Prepare working standards daily or as required by serial dilution to a range of 0 to 20 ng Hg/mL in approximately the same matrix as sample solutions. Commercially available 1,000 mg Hg/L standard solution can be used.

Digestion Procedure

1. Weigh 0.500 g dried, ground (<0.5-mm) plant tissue on a tared cigarette paper. Fold paper and contents into a ball and drop into a pre-weighed 100-mL digestion tube. If a Folin-Wu digestion tube is used, pre-weighing is not necessary.
2. Add 50 (±5) mg V_2O_5 powder followed by 5 mL HNO_3 into each digestion tube.

3. After foaming has subsided, heat tube and contents in a block digester (preheated to 160°C) for 5 minutes.
4. Remove the tube from the block digester, let cool, and then add 5 mL H_2SO_4.
5. Replace digestion tube back into the block digester and heat for an additional 15 to 20 minutes.
6. Remove the tube from the block digester, let cool, and make up to 50 mL mark with deionized water, assuming a solution density of 1.0 g/mL.

Analysis

Mercury vapor may be generated with stannous chloride ($SnCl_2$) or stannous sulfate ($SnSO_4$) (Uthe et al., 1970; van Delft and Vos, 1988) or $NaBH_4$ (Rooney, 1976; Sturman, 1985) as the reductant, and in batch or continuous-flow mode. The continuous flow analysis is more sensitive and precise than the batch mode (Dominski and Shrader, 1985). With the hydride generation technique, however, $NaBH_4$ is the only reagent required, thus reducing opportunities for Hg contamination. For this reason, and for convenience, the use of $NaBH_4$ as a reductant is favored. The reaction takes place in an acidic medium according to the equation:

$$BH_4^- + H^+ + 3H_2O \rightarrow H_3BO_3 + 4H_2$$

When using the Varian VGA-76 continuous flow vapor generation accessory, the recommended flow rates are sample/standard = 7 mL/min and $NaBH_4$ (0.3% w/v in 0.5M NaOH) = 1.1 mL/min. The acid channel may be replaced with deionized water. Nitrogen gas flow rate is 45 mL/min at a back pressure of 250 kPa. A wavelength of 253.7 nm is used for Hg analysis. Sturman (1985) reported that a flow-through absorption cell gave better results than the T-shaped cell normally used for As and Se. Sensitivity of Hg analysis is between 0.2 to 0.3 ng/mL (Dominski and Shrader, 1985; Sturman, 1985).

The reader is referred to Hamm and Stewart (1973) for the construction of a batch-type Hg vapor generator using $SnSO_4$ as the reductant.

Comments

1. Conditioning of the hydride generation system is particularly important for Hg determination. Before making measurements, the system should be conditioned by alternating the highest standard and the blank until stable results are achieved. This normally takes five to six tries.

2. Solutions of $NaBH_4$ in 0.5% NaOH are stable for at least 2 days. Deterioration of the solution is indicated by bubbles of hydrogen (H_2) collecting on the walls of the storage vessel.

3. Knechtel and Ross (1979) recommended making up Hg standard working solutions in 0.54 mM potassium dichromate ($K_2Cr_2O_7$) for enhanced stability. However, Dominski and Shrader (1985) reported that Hg standards in 0.34 mM $K_2Cr_2O_7$) caused a 5% loss of sensitivity after 24 hours. Standard solutions should be stored in glass bottles (Pyrex or soft glass).

4. Glassware should be soaked overnight in $4M$ HNO_3 before use.

5. Vanadium pentoxide may be rendered Hg-free by incineration.

6. Blanks should be determined by carrying out the complete procedure and analysis with reagents but no samples.

7. A typical rate of measurement with an autosampler in place, taking three 5-second integrations at steady state, and a 45-second delay between samples for equilibration is 60 measurements per hour.

REFERENCES

Abu-Samra, A., J.S. Morris, and S.R. Koirtyohann. 1975. Wet ashing of some biological samples in a microwave oven. *Anal. Chem.* 47:1475–1477.

Adriano, D.C. 1986. *Trace Elements in the Terrestrial Environment.* Springer-Verlag, New York.

Analytical Methods Committee. 1959. Notes on perchloric acid and its handling in analytical work. *Analyst* 84:214–216.

Barrett, P., L.J. Davidowski, Jr., K.W. Penaro, and T.R. Copeland. 1978. Microwave oven-based wet digestion technique. *Anal. Chem.* 50:1021–1023.

Beach, L.M. 1992. *Determination of As, Sb, and Se in difficult environmental samples by hydride generation.* Publication No. AA-105. Varian, Fernando, CA.

Bowen, H.J.M. 1979. *Environmental Chemistry of the Elements.* Academic Press, London, England.

Brumbaugh, W.G. and M.J. Walther. 1989. Determination of arsenic and selenium in whole fish by continuous flow hydride generation atomic absorption spectrophotometer. *J. Assoc. Off. Anal. Chem.* 72:484–486.

de Oliveira, E., J.W. McLaren, and S.S. Bergman. 1983. Simultaneous determination of arsenic, antimony, and selenium in marine samples by inductively coupled plasma atomic emission spectrometry. *Anal. Chem.* 55:2047–2050.

Deitz, F.D., J.L. Sell, and D. Bristol. 1973. Rapid, sensitive method for determination of mercury in a variety of biological samples. *J. Assoc. Off. Anal. Chem.* 56:378–382.

Dominski, P. and D.E. Shrader. 1985. *Automated cold vapour determination of mercury: EPA stannous chloride methodology.* Publication No. AA-51. Varian, Fernando, CA.

Hahn, N.H., K.A. Wolnik, F.L. Frick, and J.A. Caruso. 1982. Hydride generation/condensation system with an inductively coupled argon plasma polychromator for determination of arsenic, bismuth, germanium, selenium, and tin in foods. *Anal. Chem.* 54:1048–1052.

Hamm, J.W. and J.W.B. Stewart. 1973. A simplified procedure for the determination of total mercury in soils. *Commun. Soil Sci. Plant Anal.* 4:233–240.

Huang, J., D. Goltz, and F. Smith. 1988. A microwave dissolution technique for the determination of arsenic in soils. *Talanta* 35:907–908.

Jacobs, L.W., D.R. Keeney, and L.M. Walsh. 1970. Arsenic residue toxicity to vegetable crops grown on Plainfield sand. *Agron. J.* 62:588–591.

Kabata-Pendias, A. and H. Pendias. 1995. *Trace Elements in Soils and Plants.* Revised edition. CRC Press, Boca Raton, FL.

Knechtel, J.R. and J.L. Fraser. 1979. Wet digestion method for the determination of mercury in biological and environmental samples. *Anal. Chem.* 51:315–317.

Malaiyandi, M. and J.P. Barrette. 1972. Wet oxidation method for the determination of submicrogram quantities of mercury in cereal grains. *J. Assoc. Off. Anal. Chem.* 55:951–959.

Merry, R.H. and B.A. Zarcinas. 1980. Spectrophotometric determination of arsenic and antimony by the silver diethyldithiocarbamate method. *Analyst* 105:558–563.

National Research Council Canada (NRCC). 1979. *Effect of Mercury in the Canadian Environment.* NRCC No. 16739. Ottawa, Canada.

Nielsen, F.H. 1984. Ultra-trace elements in nutrition. *Annu. Rev. Nutr.* 4:21–41.

Nygaard, D.D. and J.H. Lowry. 1982. Sample digestion procedures for simultaneous determination of arsenic, antimony, and selenium by inductively coupled argon plasma emission spectrometry with hydride generation. *Anal. Chem.* 54:803–807.

Olson, K.W., W.J. Haas, Jr., and V.A. Fassel. 1977. Multielement detection limits and sample nebulization efficiencies of an improved ultrasonic nebulizer and a conventional pneumatic nebulizer in inductively couple plasma-atomic emission spectrometry. *Anal. Chem.* 49:632–637.

Pais, I. and J.B. Jones, Jr. 1996. *Handbook on Trace Elements in the Environment.* St. Lucie Press, Boca Raton, FL.

Pretorius, L., P.L. Kempster, H.R. van Vliet, and J.F. van Staden. 1992. Simultaneous determination of arsenic, selenium, and antimony in water by an inductively coupled plasma hydride method. *Fresenius Z. Anal. Chem.* 352:391–393.

Rooney, R.C. 1976. Use of sodium borohydride for cold vapour atomic absorption determination of trace amounts of inorganic mercury. *Analyst* 101:678–682.

Small, H.G., Jr., and C.B. McCants. 1961. Determination of arsenic in flue-cured tobacco and in soils. *Soil Sci. Soc. Am. Proc.* 25:346–348.

Smith, G.F. 1953. The wet ashing of organic matter employing hot concentrated perchloric acid. *Anal. Chim. Acta* 8:397–421.

Sturman, B.T. 1985. Development of a continuous flow hydride and mercury vapour generation accessory for atomic absorption spectrophotometry. *Appl. Spectrosc.* 39:48–56.

Thompson, A.I. and P.A. Thoresby. 1977. Determination of arsenic in soil and plant materials by atomic absorption spectrophotometry with electrothermal atomization. *Analyst* 102:9–16.

Uthe, J.F., F.A.J. Armstrong, and M.P. Stainton. 1970. Mercury determination in fish samples by wet digestion and flameless atomic absorption spectrophotometry. *J. Fish. Res. Bd. Canada* 27:805–811.

Uthe, J.F., H.C. Freeman, J.R. Johnston, and P. Michalik. 1974. Comparison of wet ashing and dry ashing for the determination of arsenic in marine organisms, using methylated arsenicals for standards. *J. Assoc. Off. Anal. Chem.* 57:1363–1365.

van Delft, W. and G. Vos. 1988. Comparison of digestion procedures for the determination of mercury in soils by cold vapour atomic absorption spectrometry. *Anal. Chim. Acta* 209:147– 156.

Zasoski, R.J. and R.G. Burau. 1977. A rapid nitric-perchloric acid digestion method for multielement tissue analysis. *Commun. Soil Sci. Plant Anal.* 8:425–436.

Zhu, B. and M.A. Tabatabai. 1995. An alkaline oxidation method for determination of total arsenic and selenium in sewage sludges. *J. Environ. Qual.* 24:622–626.

DETERMINATION OF CADMIUM, CHROMIUM, COBALT, LEAD, AND NICKEL IN PLANT TISSUE

24

Yoong K. Soon

INTRODUCTION

The method described is for the determination in plant tissue of cadmium (Cd), chromium (Cr), cobalt (Co), lead (Pb), and nickel (Ni), elements that are found at relatively low concentrations (<10 mg/kg) (Kabata-Pendias and Pendias, 1995; Pais and Jones, 1996). The method chosen for sample preparation and analysis can have a significant effect on the obtained assay results. Therefore, care needs to be exercised when preparing and assaying plant tissue for these five elements.

ORGANIC MATTER DESTRUCTION METHODS

Organic materials in plant tissues are readily destroyed by either dry or wet oxidation procedures. Problems associated with methods for the destruction of organic matter prior to metal analysis have been the subject of many papers and are most thoroughly covered in the books by Gorsuch (1970) and Bock (1978) and review articles by Gorsuch (1976) and Tolg (1974). Dry ashing is simple,

1-57444-124-8/98/$0.00+$.50
© 1998 by CRC Press LLC

requires only a mineral acid solution to dissolve the ash, and is suitable for large numbers of samples. The commonly preferred temperature of dry ashing is between 450 to 500°C, and the duration from 4 to 18 hours (McLean, 1976; Isaac and Jones, 1972). Wet acid digestion, on the other hand, requires considerable amounts of reagents and more operator attention than dry ashing. Although both methods show good agreement for a range of elements and can be used with equal success provided sufficient attention is given to procedural details and consideration of sample types (Giron, 1973; Isaac and Johnson, 1975; Watling and Wardale, 1979), the preferred method of sample preparation for the five metals covered in this chapter is dry ashing (Brooks et al., 1977; Dahlquist and Knoll, 1978; Blincoe et al., 1987). A lower final solution:sample ratio obtainable with dry ashing is also advantageous for the analysis of the trace elements. Wet acid digestion with nitric acid (HNO_3) alone gives good recovery of Ni but questionable recovery for Cr (Blincoe et al., 1987). Abu-Samra et al. (1975), however, found that a microwave oven-based wet acid digestion gave satisfactory recoveries for Co, Cr, Ni, and Pb in two reference materials, while Huang and Schulte (1985) and Rechcigl and Payne (1990) found that HNO_3 digestion followed by inductively coupled plasma emission spectrometry (ICP-AES) analysis gave excellent results for both plant macro- and micro-elements. Further investigation of these wet acid digestion-ICP-AES procedures, by including the above-mentioned five metals and other toxic metals, is warranted (Munter et al., 1984). A double dry ashing procedure gave better recoveries of Cr and Ni compared to a single ashing procedure (Dahlquist and Knoll, 1978; Blincoe et al., 1987). Therefore, a double dry ashing technique is recommended, especially when Cr analysis is required.

The procedure described here also differs from that recommended for plant macro- and micro-elements in that a narrower sample:final solution ratio is required to facilitate the detection of sub-microgram quantities of the above-mentioned elements commonly found in plant tissue samples by either atomic absorption spectrometry (AAS) (Watson and Isaac, 1990) or ICP-AES (Munter and Grande, 1981).

EQUIPMENT

1. Tall-form glazed porcelain crucibles, 30-mL capacity.
2. Analytical balance (accurate to 1.0 mg).
3. Volumetric flasks, 10-mL capacity.
4. Electric hot plate.
5. Repipette dispensers.

6. Muffle furnace.
7. Atomic absorption spectrophotometer or inductively coupled plasma emission spectrophotometer.

Reagents

1. Nitric acid (HNO_3), 70%, BDH Analar grade or better.
2. Calibration Standards (Cd, Co, Cr, Ni, and Pb in 5% HNO_3): prepare from commercially available 1,000 mg/L standards by serial dilutions. See guidelines (below) for preparing multielement standards.

Dry Ash Oxidation procedure

1. Weigh 1.000 g dried, ground plant material (<0.8 mm) previously dried at 65°C for 2 hours into a clean acid-washed crucible and place in a muffle furnace.
2. Ash at 480°C overnight (16 hours). Remove the crucible from the muffle furnace.
3. When cool, add 10 drops of deionized water, and then carefully add 2 mL 50% (v/v) HNO_3.
4. Evaporate to dryness on a hot plate (~100°C).
5. Return the crucible to the muffle furnace (cooled to <200°C) and ash at 450°C for 30 min. Remove the crucible from the muffle furnace.
6. When cool, dissolve the ash in 2 mL 20% (v/v) HNO_3 by heating on a hot plate at approximately 100°C.
7. Transfer the digest quantitatively to a 10-mL volumetric flask and dilute to volume with deionized water. If necessary, filter contents through a Whatman No. 42 filter and collect the filtrate in a plastic vial taking care to discard the first 2 mL (approximately) of filtrate.
8. Store samples at or below 10°C and analyze within a week. Loss of solution by evaporation is minimized by the low temperature.

ANALYSIS PROCEDURES

Plant tissues normally contain, on a dry mass basis, 20 to 100 µg Cd/kg; 30 to 1,000 µg Co/kg; 0.5 to 2 mg Cr/kg; 1 to 5 mg Ni/kg; and 3 to 20 mg Pb/kg (Arthur et al., 1953; McLean, 1976; Dahlquist and Knoll, 1978; Kabata-Pendias and Pendias, 1995; Pais and Jones, 1996). Hyperaccumulator species have concentrations that are up to several hundred times greater than these levels (Brooks et al., 1977). As a general guide, if all five elements are required to be analyzed, the analyst should prepare multielement standards in 0.5M HNO_3 containing Cd, Co, Cr, Ni, and Pb in the ratio 1:1:10:10:50. The exact ratio will vary with the source and nature of the plant material and the chemistry of the soil or growing medium. Further adjustments in the ratio may be made with experience gained from analyzing particular types of samples. The preferred method of analysis is ICP-AES because of its sensitivity, relative freedom from interferences, and simultaneous multielemental capability (Watson and Isaac, 1990). The normal recommended wavelengths (in nm) and detection limits (in parentheses) are Co, 238.9 (0.3 μg/L); Cr, 283.6 (0.2 μg/L); Cd, 226.5 (0.1 μg/ L); Ni, 231.6 (0.5 μg/L); and Pb, 220.3 (1.5 μg/L).

For laboratories that do not have access to an ICP-AES spectrometer, it is often possible to quantitatively determine Cd, Ni, and Pb by air-acetylene flame AAS by using a 1.0-g sample in a final volume of 10 mL with the use of slotted or concentrator tubes (Watling and Wardale, 1979). In this case, Co and prob-ably Cr should be determined by electrothermal atomization. Hoenig and de Borger (1983) considered Cr analysis of plant digests by a nitrous oxide-acety-lene flame to be adequate for most routine work. Since Cr(III) is the most common form of Cr (Adriano, 1986), it would be preferable to use Cr(III) solutions for calibration.

COMMENTS

1. The crucibles and volumetric flasks to be used should be soaked overnight in 20% v/v HNO_3 before use.
2. Greweling (1976) noted that complete oxidation of carbon is unnecessary in most circumstances and Step 5 may be omitted. The intermediate treatment with HNO_3 is, however, advised when Cr analysis is required (Dahlquist and Knoll, 1978; Blincoe et al., 1987). Munter and Grande (1981) found that heating the dissolving acid enhanced the release of some elements from the ash.
3. The overnight ashing (16 to 18 hours) is recommended as a matter of convenience. Ashing at 500°C for 4 hours has been found to be suitable for most metals (Isaac and Jones, 1972; Blincoe et al., 1987). It is recom-mended that accuracy for setting the muffle furnace temperature be ±15°C or better.

References

Abu-Samra, A., J.S. Morris, and S.R. Koirtyohann. 1975. Wet ashings of some biological samples in a microwave oven. *Anal. Chem.* 47:1475-1477.

Adriano, D.C. 1986. *Trace Elements in the Terrestrial Environment.* Springer-Verlag, New York.

Arthur, D., I. Motzak, and H.D. Branton. 1953. The determination of Co in forage crops. *Can. J. Agr. Sci.* 33:1–15.

Blincoe, C., M.O. Thiesen, and K. Stoddard-Gilbert. 1987. Sample oxidation procedures for the determination of chromium and nickel in biological material. *Commun. Soil Sci. Plant Anal.* 1 8:687–697.

Bock, R. 1978. *Handbook of Decomposition Methods in Analytical Chemistry.* International Textbook Company, Glasgow, Scotland.

Brooks, R.R., E.D. Wither, and B. Zepernick. 1977. Cobalt and nickel in *Rinorea* species. *Plant Soil* 47:707–712.

Dahlquist, R.L. and J.W. Knoll. 1978. Inductively coupled plasma-atomic emission spectrometry: Analysis of biological materials and soils for major, trace and ultratrace elements. *Appl. Spectrosc.* 32:1–29.

Giron, H.C. 1973. Comparison between dry ashing and wet digestion in the preparation of plant material for atomic absorption analysis. *Atomic Absorption Newsletter* 12:28–29.

Gorsuch, T.T. 1970. *The Destruction of Organic Matter.* Pergamon Press, Oxford, England.

Gorsuch, T.T. 1976. Dissolution of organic matter, pp. 491–508. In: P.D. LaFluer (Ed.), *Accuracy in Trace Analysis, Sampling, Sample Handling, Analysis.* Volume 1. Special Publication 422. National Bureau of Standards, Washington, D.C.

Greweling, T. 1976. Chemical analysis of plant tissue. Agronomy No. 6. Search Agriculture 6(8):1–35. Cornell University Agricultural Experiment Station, Ithaca, NY.

Hoenig, M. and R. de Borger. 1983. Particular problems encountered in trace metal analysis of plant material by atomic absorption spectrometry. *Spectrochim. Acta* 38B:873–880.

Huang, C.Y.L. and E.E. Schulte. 1985. Digestion of plant tissue for analysis by ICP emission spectroscopy. *Commun. Soil Sci. Plant Anal.* 16:943–958.

Isaac, R.A. and J.B. Jones, Jr. 1972. Effects of various dry ashing temperatures on the determination of 13 nutrient elements in five plant species. *Commun. Soil. Sci. Plant Anal.* 3:261–269.

Isaac, R.A. and W.C. Johnson. 1975. Collaborative study of wet and dry ashing techniques for the elemental analysis of plant tissue by atomic absorption spectrophotometry. *J. Assoc. Off. Anal. Chem.* 58:436–440.

Kabata-Pendias, A. and H. Pendias. 1995. *Trace Elements in Soils and Plants.* Second edition. CRC Press, Boca Raton, FL.

McLean, A.J. 1976. Cadmium in different plant species and its availability in soils as influenced by organic matter and additions of lime, P, Cd, and Zn. *Can. J. Soil Sci.* 56:129–138.

Munter, R.C. and R.A. Grande. 1981. Plant analysis and soil extract analysis by ICP-atomic emission spectrometry, pp. 653–672. In: R.M. Barnes (Ed.), *Developments in Atomic Plasma Spectrochemical Analysis*. Heyden and Son, Ltd., London, England.

Munter, R.C., T.L. Halverson, and R.D. Anderson. 1984. Quality assurance of plant tissue analysis by ICP-AES. *Commun. Soil Sci. Plant Anal*. 15:1285–1322.

Pais, I. and J.B. Jones, Jr. 1996. *Handbook on Trace Elements in the Environment*. St. Lucie Press, Boca Raton, FL.

Rechcigl, J.E. and G.G. Payne. 1990. Comparison of a microwave digestion system to other digestion methods for plant tissue analysis. *Commun. Soil Sci. Plant Anal.* 21:2209–2218.

Tolg, G. 1974. The basis of trace analysis, pp. 698–710. In: E. Korte (Ed.), *Methodium Chimicuin*. Volume 1. Analytical Methods. Part B. Micromethods, Biological Methods, Quality Control, Automation. Academic Press, New York.

Watling, H.R. and I.M. Wardale. 1979. Comparison of wet and dry ashing for the analysis of biological materials by atomic absorption spectroscopy, pp. 69–80. In: L.R.P. Butler (Ed.), *The Analysis of Biological Materials*. Pergamon Press, Oxford, England.

Watson, M.E. and R.A. Isaac. 1990. Analytical instruments for soil and plant analysis, pp. 691–740. In: R.L. Westerman (Ed.), *Soil Testing and Plant Analysis*. Third edition. SSSA Book Series 3. Soil Science Society of America, Madison, WI.

QUALITY CONTROL PROCEDURES FOR PLANT ANALYSIS

25

Edward A. Hanlon

INTRODUCTION

This section describes a system of Quality Assurance (QA) methods that apply Total Quality Management (TQM) principles specifically to plant analyses. The intent of TQM is to incorporate quality into every aspect of the system to produce a high-quality product the first time. This approach acknowledges that everyone desires to perform their work well. In fact, most problems are not caused by people, but by forcing people to work within a process that precludes them from performing well. It is the manager's responsibility to change the process, and the technician's lot to work within that process. Employees are empowered when they have the statistical tools to evaluate their work. Also, management can encourage process improvement simply by asking those who work within the process to identify stumbling blocks to quality. Constant refinement produces a high-quality process.

EQUIPMENT

1. A computer with spreadsheet software, graphics, and statistical capability is highly desirable.

1-57444-124-8/98/$0.00+$.50
© 1998 by CRC Press LLC

PROCEDURE

1. Construct a customer-supplier diagram for each process within the plant
   tissue laboratory. The following diagram (Figure 25.1) describes a generic
   plant tissue analysis process. The diagram is not intended to be complete,
   only to show the use of a customer-supplier diagram (Hanlon, 1992, 1996).
   The diagram allows review of each process step and shows who is responsible. Each step of the real laboratory process must appear in the diagram
   to be of use. Development of this type of diagram is a team project including all of those involved in the process.

   A customer-supplier relationship exists between each connected column.
   For example, Technician 1 accepts sample custody from the client, passes
   payment to the Secretary and client information to the Data Manager. This
   diagram shows that Technician 1 is in direct contact with the client, and
   handles funds and sample custody.

   Each of these actions crosses the columns within the diagram indicating
   separate customer-supplier relationships. Interaction with the customer is
   important because (1) the customer must supply samples suitable for analytical work; (2) the client's impression of the laboratory is affected by this
   interaction with Technician 1; and (3) quality must start with this inter-
   change (correct sample information, sample condition, payment, etc.).

   Technician 1 is a supplier to both the Secretary and the Data Manager.
   They in turn are customers of Technician 1. At each customer-supplier
   interface, quality must be reviewed and improved. How can the Secretary
   prepare an accurate bank deposit if Technician 1 incorrectly identifies sample
   payment information? If the Secretary is supplied with poor quality information, the laboratory must pay for either Technician 1 or the Secretary to
   redo this portion of the process.
2. Prepare a Standard Operating Procedure (SOP) manual. Each of the activities within the customer-supplier diagram must appear in the SOP. The
   specific steps with applicable techniques and procedures should be explained
   in sufficient detail to act as a reference for experienced personnel, a training
   manual for novices, and a detailed method source for inspectors.
3. Analyze external references and standards every time unknown plant samples
   are analyzed. External references and standards are plant tissue and solutions with certified concentrations prepared by commercial suppliers. Accuracy of a laboratory procedure is documented by analyzing these external
   samples and solutions. Results from these references and standards form
   one level of the QA plan of the laboratory (Taylor, 1989).

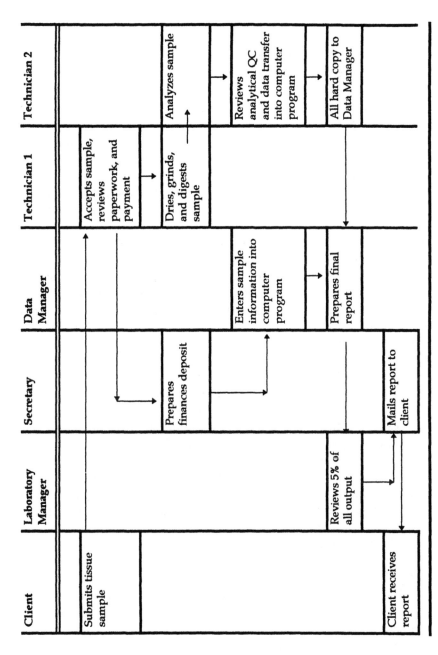

FIGURE 25.1   Customer-supplier diagram.

4. Develop internal references and standards from local sources and chemicals to augment external sources and standards every time unknown plant tissue is analyzed. Internal references and standards consist of plant tissue and solutions that have been prepared within the laboratory. The precision of the laboratory procedure can be documented by these internal samples and solutions. Results from these references and standards form a second level of the QA plan of the laboratory (Taylor, 1989).
5. Replications of unknown samples form a third level of the QA plan. Replications may be completed through the entire process (i.e., preparation, digestion, analysis), or through specific portions (i.e., duplicate analyses of the same digestion).
6. Spiked unknown samples for tissue are often difficult to produce. Spiking the digested solution is a more reliable way of demonstrating recovery and the accuracy of the procedure.
7. Quality Control Charts should be used to determine the quality of data being generated by the laboratory. These charts are meant to be used during the analysis process to allow the technician to stop analyzing if the process is not in statistical control. Therefore, use of external and internal standard tissue samples and solutions, replications, and spiked sample solutions must be incorporated within the analytical stream, not at the end of the analysis. The trained technician uses these tools to ensure that quality work is done the first time.
   a. **X or Run Chart** (Figure 25.2; after Hanlon, 1996): This chart shows changes in a specific sample (such as an external plant sample) with time. The mean value and both upper and lower warning and control lines are added to the graph to aid the technician with interpretation of each reading as it is made (Gitlow et al., 1989).

      The warning and control lines are calculated from a minimum of 15 observations. These calculations should be made after any major improvements to the process are made. Ideally, the old warning and control lines should be wider than the newly calculated lines. This change is indicative of a process that has been improved.
      - Mean or central line = average of observations
      - Warning lines = ±1 to 2× standard deviation of the observations
      - Control lines = ±3× standard deviation of the observations

      When one observation falls outside of the control lines, or two consecutive observations lie between the warning and control lines, the process should be recalibrated, and all unknown samples since the last acceptable observation should be re-analyzed (Taylor, 1989). Figure 25.2 shows an X Chart for copper (Cu) in a National Institute of Standards

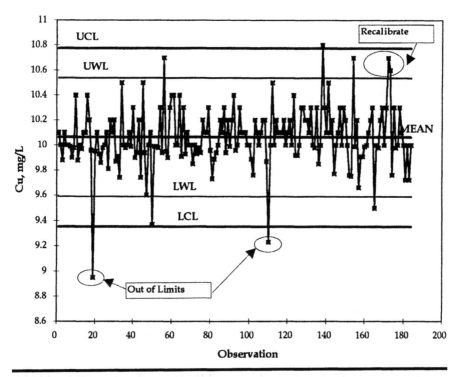

FIGURE 25.2 Example of an X Chart. Each point is the observed value of the Cu standard; LCL = lower control limit; LWL = lower warning limit; Mean = arithmetic mean of observations; UWL = upper warning limit; and UCL = upper control limit.

and Technology tissue sample. Note that there are several observations that caused the process to be recalibrated.

b. **R Chart** (Figure 25.3; after Hanlon, 1996). This chart is useful in determining the precision of a process. The plotted points are the difference between two or more readings of the same sample. Decisions concerning recalibration are the same as for the X Chart. An R Chart contains only a mean and upper warning and control lines.

Unlike the X Chart, an R Chart does not use the same sample with time. R Charts often are developed for different concentration ranges. Limiting the concentration range also reduces the possibility of comparing replicates with low concentration with those of high concentration. For example, Figure 25.3 contains differences between replicate analyses whose concentration ranged from 0 to 15 mg/L. Note that during this

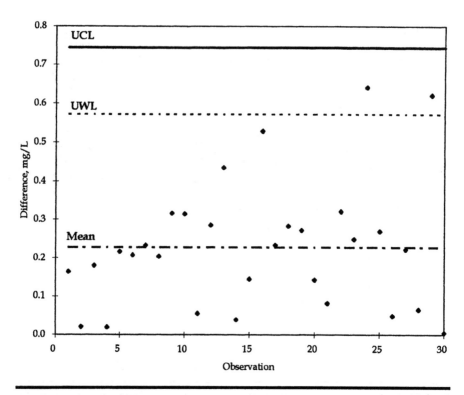

FIGURE 25.3 Example of an R Chart. Each point is the arithmetic difference between two observations of the same sample; UWL = upper warning limit; UCL = upper control limit; and Mean = arithmetic mean of differences.

month the process was in statistical control the entire time (i.e., all replicate differences were acceptable).

c. **Pareto Chart** (Figure 25.4; after Hanlon, 1996). The Pareto Chart is best used as a problem-solving tool. After statistical testing has identified problems, the Pareto Chart organizes findings in order of occurrence. The frequency of problem occurrence decreases from left to right. Therefore, focusing problem-solving skills of a team on these frequently encountered problems will produce the most benefit or improvement to the system.

While the Pareto Chart can be a simple histogram, it is often more useful when considering two or three factors together. Figure 25.4 presents the in-lab time for a number of processes compiled for a 6-month period in 1992. The process with the longest in-lab time (7.5 days) is

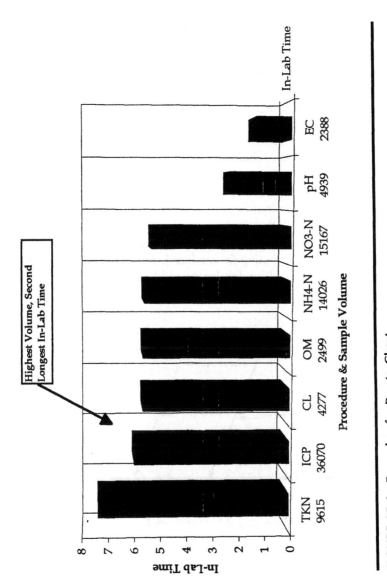

FIGURE 25.4   Example of a Pareto Chart.

TKN. Using a second factor, number of samples, there are 3.75 times the number of samples for TKN being analyzed through the ICP process with an average in-lab time of 5.9 days. Obviously, reduction of in-lab time for the ICP process will affect a larger number of samples (and clients) than a similar reduction in the TKN process. Team effort addressed this problem with the ICP reducing the in-lab time to about 4.5 days with a sample volume increase to greater than 50,000 for the same 6-month period the next year.

d. **Cause-and-Effect Diagram** (Figure 25.5). Another problem-solving tool for tissue analyses is the development of a Cause-and-Effect Diagram. The diagram is a graphical listing of all known or suspected sources of variability within the selected process. Figure 25.5 includes major categories (equipment, process, data control, sample, supplies, and personnel). These categories suit tissue analysis quite well, but are somewhat different from those traditionally used in these diagrams (Gitlow et al., 1989; Weaver, 1991).

Contributing factors are listed within each category. While Figure 25.5 has some contributing factors listed, the list is not exhaustive. The recommended method for developing and listing all contributing factors

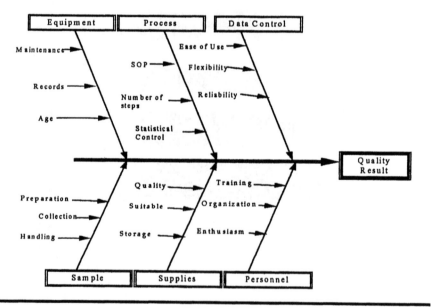

FIGURE 25.5   Example of a Cause-and-Effect Diagram.

is through the use of a team whose members are familiar with the targeted process.

After development of the Cause-and-Effect Diagram, individual contributing factors can be tested and modified, progressively refining the process to remove variability. The diagram can often be used as a quick reference when trying to diagnose exceptional causes preventing the process from being in statistical control.

REFERENCES

Gitlow, H., S. Gitlow, A. Oppenheim, and R. Oppenheim. 1989. *Tools and Methods for the Improvement of Quality*. Irwin, Homewood, IL.

Hanlon, E.A. 1992. Quality assurance plans for agricultural testing laboratories, pp. 18-32. In: J.B. Jones, Jr. (Ed.), *Handbook on Reference Methods for Soil Analysis*. Soil and Plant Analysis Council, Athens, GA.

Hanlon, E.A. 1996. Laboratory quality: A method for change. *Commun. Soil Sci. Plant Anal.* 27:307-325.

Taylor, J.K. 1989. *Quality Assurance of Chemical Measurements*. Lewis Publishers, Chelsea, MI.

Weaver, C.N. 1991. *TQM: A Step-By-Step Guide To Implementation*. American Society of Quality Control, Milwaukee, WI.

REFERENCE MATERIALS FOR DATA QUALITY CONTROL

26

Milan Ihnat

INTRODUCTION

Reliable analytical measurements are mandatory for legal compliance with government regulations, long-term monitoring and baseline studies, standardization of laboratory performances, and research. In agriculture/food science, accurate data on the chemical composition of raw agricultural products and foods are needed to assess effects of farm management practices, and of changes in crop culture and food processing on nutrient and toxic chemical content of retail food products. Elemental concentration information is required to establish the essentiality of a nutrient or toxicology of a toxicant to determine the roles of nutrients in health and disease, to identify adequate, subadequate, or marginal intakes by the population, to establish nutrient dietary requirements, to accumulate baseline concentration data in order to investigate the effects on nutrient levels of various methods for food processing, and to comply with legal labeling requirements.

Reasons for the general lack of agreement among laboratories of the outcome of analytical work stem from a multitude of factors inherent in every analysis, contributing errors to the final measurements, broadly categorized as presampling, sampling, sample manipulation, and measurement factors that include other considerations such as contamination, data quality control, and analyst competence.

1-57444-124-8/98/$0.00+$.50

The incorporation of appropriate Reference Materials into the analysis scheme utilizing good methods and other aspects of a quality control program is the most convenient, cost-effective mechanism by which to assess, monitor, and maintain analytical data quality and ensure accuracy of results. Although this chapter is closely tied in to the chapter on quality assurance/quality control (QA/QC), and in fact the Reference Material concept may be considered a subset of the larger topic of QA/QC, the subject of Reference Materials is of sufficient magnitude to stand alone as it is an integral component of the principles of plant analysis.

In this chapter, the concept and role of Reference Materials are summarized and procedures for their selection and utilization are presented to assist the plant scientist and analyst in monitoring and maintaining analytical data quality in the determination of inorganic analytes. Reference is made to tables of plant and related Reference Materials presented in Appendix I, listing available products, sources of supply, and some common elemental concentrations as guides to reference material selection. The thrust of this chapter is to describe control procedures for the analysis of inorganic chemical composition utilizing Reference Materials. Furthermore, while extractable or bioavailable elemental concentrations are frequently of interest, only the control of total elemental concentrations will be addressed because of the usual requirement in plant analyses for total contents and the lack of Reference Materials for extractable analytes. The concepts and methods of application summarized here apply to all methods and elements reported in this manual insofar as the analytes are represented by existing Reference Materials. For completeness, elements other than those dealt with in this manual are listed for future use as research and method development are extended to elements other than those of current interest to plant scientists.

NATURE AND ROLE OF REFERENCE MATERIALS

A Reference Material is considered to be any material, device, or physical system for which definitive numerical values can be associated with specific properties and that is used to calibrate a measurement process. The term Reference Materials (Uriano and Gravatt, 1977) is used to describe a generic class of well-characterized, stable, homogeneous materials produced in quantity and having one or more physical or chemical properties experimentally determined within stated measurement uncertainties. Primary Reference Materials are defined as those having properties certified by a recognized standard laboratory or standards agency, such as the National Institute of Standards and Technology (NIST). Primary Reference Materials produced by NIST are denoted Standard Reference Materials (SRM), a term used synonymously with the term Certified Reference Materials (CRM) recognized by the International Union of Pure and Applied Chemistry (IUPAC) and the International Organization for Standardization (ISO). Uriano and Gravatt (1977) further state that a primary Reference Material is normally produced by a national standards laboratory or other organization having legal authority to issue such materials. A formal definition of a NIST Standard Reference Material is as follows (Cali et al., 1975):

> NIST Standard Reference Materials (SRM) are Certified Reference Materials issued by NIST. These are well-characterized materials produced in quantity to improve measurement science. SRM's are certified for specific chemical or physical properties and are issued by NIST with certificates that report the results of the characterization and indicate the intended use of the material. They are prepared and used for three main purposes: (1) to help develop accurate methods of analysis (reference methods); (2) to calibrate measurement systems used to: (a) facilitate exchange of goods, (b) institute quality control, (c) determine performance characteristics, or (d) measure a property at the state-of-the-art limit; and (3) to assure the long-term adequacy and integrity of measurement quality assurance programs.

According to the ISO, a Reference Material is defined as "a material or substance one or more properties of which are sufficiently well established to be used for the calibration of an apparatus, the assessment of a measurement method, or for assigning values to materials." A CRM is defined as "a Reference Material, one or more of whose property values are certified by a technically valid procedure, accompanied by or traceable to a certificate or other documentation which is issued by a certifying body."

In this chapter, only Reference Materials for chemical composition quality control are considered—the term Reference Material refers to any *bona fide,* credible products with certified, recommended or best-estimate concentration values. Further, only natural matrix products, generally with native elemental contents, are discussed. Artificially prepared or spiked materials as well as pure elements for preparing calibration solutions and such available solutions are not included in this presentation.

Reference Materials are one of the most popular of the three usual mechanisms (reference data, Reference Materials, and reference methods) for achieving compatibility and transferring accuracy among laboratories (Uriano and Gravatt, 1977; Coleman, 1980). Incorporation of appropriate Reference Materials into the scheme of analysis that utilizes good methods and other aspects of a quality control program is the most convenient, cost-effective mechanism by which to assess and maintain analytical data quality. Coverage of the roles and uses of Reference Materials is discussed by Cali et al. (1975), Uriano and Gravatt (1977), Taylor (1993), and Ihnat (1988, 1993).

PRELIMINARY REQUIREMENTS

In order to produce scientifically valid analytical data and to properly and cost effectively use Reference Materials, it is imperative that compliance with several prerequisites be established.

1. An appropriate analytical method must be applied to the task on hand, by appropriately qualified and trained personnel in a suitable physical and administrative environment. Suitable physical environment refers to the equipment, materials, reagents, and laboratory conditions necessary for the proper execution of the method; suitable administrative environment includes understanding of and support for appropriate data quality by the analyst's supervisor and all other managers. The role of the analyst is of direct paramount importance; good analysis and good analyst go hand in hand. Analyst training, experience, familiarity with the problem on hand, skill, attitude, motivation, and judgment are necessary prerequisites with which satisfactory solution of analytical problems is possible.
2. Suitable quality control/quality assurance procedures should be routinely in use and the need for appropriately reliable analytical information must be recognized. The analytical system must be in a state of statistical control, i.e., operating optimally and consistently generating acceptable data. Typically, the method under test should give a precision (standard deviation) with the Reference Material and other homogeneous materials equal to or better than the uncertainty reported for the Reference Material in the certificate (Okamoto and Fuwa, 1984).
3. When dealing with the determination of total concentrations of elements, i.e., the sum of all the element concentrations in all material (sample) phases and molecular species, it must be ascertained that the method is in fact measuring all of the element. The sample decomposition procedure must bring into solution all of the material; no grains or insoluble fraction must be left behind (Ihnat, 1982). In addition, the element must be in the correct oxidation state required by the procedure. Extractable/bioavailable levels of elements are commented on later.

PROCEDURES FOR REFERENCE MATERIAL SELECTION AND USE

Selection

To facilitate application of selection and use procedures described here and to provide guidance to the analyst, detailed Reference Material information is presented in tables in Appendix I, listing suppliers of Reference Materials, products for plant and related materials, and elemental concentration values. Provided by different agencies, following different preparation and certification approaches, all Reference Materials are not to be expected to adhere to uniform criteria of quality. They are, however, believed to be good choices for routine quality control.

For correct and effective use of a Reference Material, the material selected must be appropriate to the task. The material must resemble, as closely as possible in all respects, the actual materials being analyzed. It must be very similar with respect to matrix (all constituents other than the analyte) and must contain the analyte at a concentration level and form (e.g., native form, speciation, etc.) similar to the commodity undergoing analysis. Furthermore, the Reference Material must be sufficiently homogeneous so that test portions of size commensurate with the analytical method can be used. Ideally (but often impossible), two or three materials should be chosen to bracket the analyte composition of the sample. The following steps could be followed to select appropriate Reference Materials:

1. Select a plant Reference Material by consulting Appendix I Table 3 for materials that have given concentration values equal to or similar to that expected in the test material for the element of interest. Using the descriptive name of the product given in Appendix I Table 2, select the material approximating the laboratory sample to be controlled with respect to general type (i.e., matrix, based on name) as well as the analyte level expected. Refer to Appendix I Table 3 to independently select Reference Material(s) based on concordance of the concentration level of the element of interest to the level anticipated in the sample. Should the Reference Material as chosen in this manner from Appendix I Table 3 not be sufficiently close in matrix match, expand the concentration range considered to establish whether additional, suitable materials are netted in the larger search. If desired, fine-tune the choice with respect to matrix by consulting Appendix I Table 3 for concentrations of matrix elements (i.e., those elements constituting the bulk of the material whether or not they need to be determined). Carry out this

iterative selection process (referring to Appendix I Table 2 for preliminary material selection, Appendix I Table 3 for analyte concentration matching, and for matrix element concentration considerations) to lead to the final selection of an appropriate Reference Material.

2. Follow the same approach to choose, if possible, a second or third Reference Material, of similar matrix, approximating the analytical samples, to match (or bracket) the sample with respect to concentration of the given element.

3. For multielement analyses (the determination of more than one element on the same laboratory subsample), go through the identical material selection steps for the subsequent elements to choose appropriate materials for each of these respective analytes. Maximize the number of elements to monitor by a given Reference Material, i.e., minimize the number of materials required, by reducing the strictness of matrix and analyte matching criteria. Rigidity of selection criteria is at the analyst's discretion and is governed by the level of quality control desired, the number and availability of Reference Materials as well as the rate of Reference Material usage acceptable (refer to Notes 1 to 3).

Use

Major uses of Reference Materials within the measurement process are generally: (1) analytical calibration, (2) quality control, (3) analytical method development and evaluation, and (4) production and evaluation of other reference materials (Cali et al., 1975; Uriano and Gravatt, 1977; Taylor, 1993). Utilization of natural matrix Reference Materials for establishing calibration functions (analytical response as a function of analyte concentration) (Morrison, 1975) is not generally recommended due to uncertainties in certified elemental concentrations. Such uncertainties, resulting from material nonhomogeneity and certification measurement imprecisions and biases, are generally several-fold greater than the compositional uncertainties for pure elements or compounds usually used for calibration solutions. Since the concentrations of standard solutions used for calibration should be known with greater certainty than is required in the analysis of an actual sample, the use of high purity, highly reliable pure elements and compounds rather than natural matrix Reference Materials are preferred for calibration.

The recommended mode of Reference Material usage is for analytical data quality control to establish method performance (bias) and to monitor and maintain data quality (Taylor, 1993). Errors in measurement can arise in the three component steps of an analytical method: (1) sampling (including presampling considerations), (2) sample manipulation, and (3) measurement. Thus, when using the Reference Material for data quality control, the aggregate of all steps subsequent to the point at which the material is introduced into the scheme of analysis will be monitored for performance. Follow the steps below for reference material utilization:

1. Ensure that the analytical system is in a state of statistical control (as stipulated under Prerequisites).
2. Following certificate/report of analysis instructions for material usage and handling, incorporate the Reference Material(s) into the scheme of analysis, at the earliest stage possible, i.e., prior to the beginning of sample decomposition. Take it through the entire analytical procedure at the same time and under the identical conditions as the actual analytical samples in order to correctly monitor all the sample manipulation and measurement steps.
3. For multielement determinations, should different sample preparation and measurement procedures (i.e., different analytical methods) be indicated for the different elements, take separate aliquots of the Reference Material through the entire relevant analytical scheme for proper quality control (refer to Notes 4 to 7).

Performance Interpretation and Corrective Action

When possible, the analysis of several Reference Materials, spanning the concentration range of interest, is the most useful way to investigate measurement bias. The three-sample approach—analysis of a low, middle, and upper range sample—is practical provided that the range of analytical interest is covered. However, acquiring the necessary Reference Materials from the world repertoire of materials may not always be possible. When supported by other data, the measurement of even a single Reference Material can be meaningful. *Standard Reference Materials: Handbook for SRM Users* (Taylor, 1993) is recommended for a detailed discussion of Reference Material use.

The method under test should usually give a precision (standard deviation) with the Reference Material equal to or better than the overall uncertainty reported for the Reference Material by the issuer (Okamoto and Fuwa, 1984). Results from the analysis of the Reference Material are then compared with the certified value; rarely will the two agree exactly due to measurement errors in each. Whether the two differ significantly is ascertained by comparing the two values, and their uncertainties using simple statistical tests. If the confidence intervals intersect, the measured concentration value agrees with the certified value, and the analyst can deduce, with some confidence, that the method is applicable to the analysis of materials of similar composition. Otherwise there is disagreement and the method as applied exhibits a bias. One of the following calculation steps can be followed to estimate agreement of the measured and certified concentration values:

1. **Case with All Parameters Available:** Compare the 95% confidence levels calculated from the standard deviation, number of analyses and the Student t statistic with the confidence or tolerance interval of the Reference Material using the following equations:

$$\bar{X}_1 - \bar{X}_2 = ts(1/n_1 + 1/n_2)^{1/2} \qquad (26.1)$$

$$s_1^2 = \frac{(n_1-1)s_1^2 + (n_2-1)s_2^2}{(n_1+n_2-2)} \qquad (26.2)$$

where $\bar{X}_1$ is the mean concentration found by the user for the Reference Material, $\bar{X}_2$ is the certified recommended or reference value for the Reference Material, s_1 is the standard deviation estimated from n_1 determinations by the user, s_2 is the standard deviation reported for the Reference Material in the certificate or report of analysis based on n_2 determinations, and t is the Student t statistic.

The difference $\bar{X}_1 - \bar{X}_2$ is compared to the right-hand side of Equation 26.1 using the t value for 95% confidence ($p = 0.05$). Should the difference be greater (positive or negative), a discrepancy exists between the measured and certified concentration values, which indicates that the analytical procedure is not operating well. Should it be ascertained that an unacceptable bias exists, a correction for it should not be applied; instead, diagnostic steps should be taken to identify sources of unacceptable bias or imprecision and corrective action should be taken to eliminate or reduce errors.

2. **Case with Missing n_2 and Negligible Uncertainty in the Reference Material Certified Value:** Compare the absolute value of the estimated bias $X_1 - X_2$ with a critical value based on:

$$\bar{X}_1 - \bar{X}_2 = \frac{ts_1}{(n_1)^{1/2}} \qquad (26.3)$$

using uncertainty parameters only for the measurements carried out by the analyst. Proceed further as in Case 1.

3. **Case with Missing n_2 and Specified Uncertainty in the Reference Material:** Compare the absolute value of the estimated bias $\bar{X}_1 - \bar{X}_2$ with a critical value based on:

$$\overline{X}_1 - \overline{X}_2 = \frac{ts_1}{(n_1)^{1/2} + u} \qquad (26.4)$$

where u is the uncertainty of the certified concentration reported in the certificate of analysis. Proceed further as in Case 1 (refer to Notes 8 to 10).

NOTES

1. The rate of incorporation of Reference Materials is at the discretion of the analyst and could range from less than one Reference Material/100 samples (more typical) to more than one per 10 samples, depending on the nature of the work and data quality requirements. Stocks of Reference Materials can be conserved by including laboratory control materials for more frequent monitoring, reserving Reference Materials for critical control. In large routine analysis operations, where many similar samples are analyzed concurrently in a batch or run, one aliquot of a suitable Reference Material will suffice to monitor the performance of the method for quite a number of samples.

2. A preliminary semiquantitative analysis of the sample would be advantageous to facilitate selection of a closely matching control material, but this usually may not be feasible unless one has access to high throughput, multielement analytical techniques or is carrying out high-reliability determinations. Selection of the Reference Material based on similarity of its matrix (crop name) with the sample may suffice.

3. For multielement analyses, it would be efficient and cost effective to be able to use the same aliquot of the selected Reference Material for quality control. Feasibility of this approach depends on whether that Reference Material has certified analytical values for the elements of interest and whether the concentration levels in the Reference Materials and submitted materials are in reasonable concordance. Suitable matching is left to the analyst's discretion and will require consideration of level of control desired, number, and availability of Reference Materials and frequency of incorporation in the analytical scheme.

4. The information in the tables in the Appendix related to this chapter should be used only as guides for Reference Material selection and use. *The latest appropriate certificates or reports of analysis or other relevant publications issued with the Reference Material must be consulted and used.* The most important source of information is the certificate or report of analysis issued with the product. This document is an integral part of the Reference Material technology as it provides the analytical (certified)

information, estimates of uncertainties, instructions for the correct use of the material, and other relevant information.

5. Reference Materials can monitor the performance of laboratory procedures subsequent to the point of introduction of the Reference Material. Activities occurring prior to this, such as sampling, preservation, storage and presampling considerations, are generally impossible to monitor by use of Reference Materials.

6. It is important that both the Reference Material and actual samples undergo identical, simultaneous handling; if feasible, the Reference Material could be submitted as a blind material to the analyst. It is also important that the Reference and actual sample analyte concentrations be reasonably close, since method performance can vary dramatically with concentration, and conclusions at one level may not be applicable to other levels.

7. Reference Materials are best used on a regular basis. Their sporadic use when trouble is suspected is legitimate, but systematic measurement within a quality control framework will generally be more informative and is highly recommended. Reference Materials may be used as the sole quality control material or they may be used in conjunction with in-house or locally produced control materials in a systematic manner in order to conserve the former.

8. It cannot be too strongly emphasized that the original, appropriate certificate or report of analysis, and not other sources of information, be consulted for correct usage of a Reference Material, analytical information, and for interpretation of results.

9. It must be emphasized that compliance with prerequisites be ensured; in particular, the measurement system should be under statistical control for the proper, successful, and cost-effective application of the Reference Material concept to quality control, and before the analytical data can be used and errors identified and corrected. Identification of causes of unacceptable bias, or precision, a necessary first step prior to corrective action, is not easy. Whenever excessive bias or imprecision is found to be present, corrective action must be taken, otherwise the measurement process will have limited usefulness.

10. There are not too many instances where the uncertainty for the Reference Material is characterized by a standard deviation, s_2, and corresponding number of determinations, n_2. Thus, Case 2 and particularly Case 3 will most often be the ones of necessity. The uncertainty, u, in Case 3 is not necessarily a standard deviation or standard error, but can reflect symmetric or asymmetric estimates of imprecision and possible systematic errors among methods used in certification.

COMMENTS AND CONCLUSIONS

Reliability and absolute confidence in the stated characteristics of Reference Materials is a basic critical criterion for their use for quality control. Thus, in this presentation, no treatment has been included of in-house, local, regional, or other "uncertified" reference products. Homogeneous in-house materials, standardized with respect to approximate analyte concentration, can be incorporated into the quality control scheme for day-to-day or more frequent monitoring of precision; certified, *bona fide* Reference Materials are reserved for monitoring analytical bias. Reference Materials certified for extractable elemental concentrations in plant products, or for that matter just about any other commodities with the exception of some soils, are unavailable. Thus, use of Reference Materials for quality control of procedures for extractable analytes documented in this volume is not possible. Certification philosophy rests on the concept of independent methodology, the application of theoretically and experimentally different measurement techniques and procedures to generate method-independent concordant results; extractable concentrations are generated by specific procedures and are thus method dependent, an idea which has to be rationalized with the fundamental method-independent concept in Reference Material certification work.

Application of methods of chemical analysis is fraught with many sources of error, and countless opportunities exist for the introduction of bias and imprecision into the final numerical results. Such systems must therefore be operated under a complete, regularly applied quality assurance program if results are to be meaningful. Reference Materials are an integral, cost-effective component of quality control to monitor system performance.

The aim of this presentation is to increase awareness of the plant analyst and scientist of the concept, role, and utility of Reference Materials in data quality. It is hoped that it will stimulate increased Reference Material use as a cornerstone of the quality control program to establish, monitor, and maintain analytical data quality in the plant analysis laboratory.

ACKNOWLEDGMENT

Contribution No. 96-20 from the Centre for Land and Biological Resources Research, Agriculture and Agri-Food Canada, Ottawa, Ontario, Canada.

REFERENCES

Cali, J.P., T.W. Mears, R.E. Michaelis, W.P. Reed, R.W. Seward, C.L. Stanley, H.T. Yolken, and H.H. Ku. 1975. The Role of Standard Reference Materials in Measurement Systems. NBS Monograph 148, National Bureau of Standards, Washington, D.C.

Coleman, R.F. 1980. Improved accuracy in automated chemistry through the use of Reference Materials. *J. Automatic Chem.* 2:183–186.

Ihnat, M. 1982. Importance of acid-insoluble residue in plant analysis for total macro and micro elements. *Commun. Soil. Sci. Plant Anal.* 13:969–979.

Ihnat, M. 1988. Biological reference materials for quality control, pp. 331–351. In: H.A. McKenzie and L.E. Smythe (Eds.), *Quantitative Trace Analysis of Biological Materials.* Elsevier Science, Amsterdam, The Netherlands.

Ihnat, M. 1993. Reference materials for data quality, pp. 247–262. In: M.R. Carter (Ed.), *Soil Sampling and Methods of Analysis.* Lewis Publishers, Boca Raton, FL.

Ihnat, M. 1997. Plant and related reference materials for data quality control of elemental content, pp. 235–284. In: Y.P. Kalra (Ed.), *Handbook of Reference Methods for Plant Analysis.* St. Lucie Press, Boca Raton, FL.

Morrison, G.H. 1975. General aspects of trace analytical methods. I. Methods of calibration in trace analysis. *Pure Appl. Chem.* 41:395–403.

Okamoto, K. and K. Fuwa. 1984. Low contamination digestion bomb method using a Teflon double vessel for biological materials. *Anal. Chem.* 56:1758–1760.

Taylor, J.K. 1993. Standard Reference Materials: Handbook for SRM Users. NIST Special Publication 260-100, National Institute of Standards and Technology, Gaithersburg, MD.

Uriano, G.A. and C.C. Gravatt. 1977. The role of reference materials and reference methods in chemical analysis. *CRC Crit. Rev. Anal. Chem.* 6:361–411.

DATA PROCESSING IN PLANT ANALYSIS

27

Rigas E. Karamanos

INTRODUCTION

A data processing system must reflect the needs, requirements, and philosophy of both fertilizer recommendations and service by a laboratory. A properly designed system not only can handle the databases upon which recommendations are based, but can also control the process by which a laboratory handles and tests plant samples and communicates with clients. A program also automates the process of evaluating test results, applying proper quality assurance/quality control protocols, making fertilizer recommendations when necessary, and preparing plant tissue test reports.

With today's computer advances, the scope and extent of a system is only limited by the user's imagination and the economics of the system development. The challenges appear to be more in the structure of the database and the philosophy and agronomic information available rather than in the development of a computerized system itself. In any event, adherence to some basic principles for choosing a system appropriate for a laboratory can help alleviate wrong choices, overexpenditure and lost effort.

The first step in establishing a computerized database is to define the "need" for a computerized Data Processing System. In selecting the appropriate system, whether it is to be bought off the shelf or developed internally, all players (essentially all the laboratory staff) should participate in a process to define the functionality and attributes of the desired system and to record expectations of the various laboratory personnel as to what the system will do for them. Hence,

1-57444-124-8/98/$0.00+$.50
© 1998 by CRC Press LLC

a series of meetings should be arranged to gather input from all involved. Often, examining the existing process of handling plant samples and the problems that might be associated with it can be a good starting point in order to arrive at the goals of a computerized data processing system.

This process will lead to the formulation of the specifications for the data processing system. Once the specifications are drawn, the system design can proceed. In documenting the program requirements, there are two main parameters that must be defined: (1) functions, a list of the required functions of the program, stated as clearly as possible so as to make the statement free of ambiguity, so that only one interpretation of each function statement is possible; and (2) attributes, a list of required attributes of the program, such as fast, user-friendly, etc. Each of the attributes must be rated as to whether it is an attribute the program must have, would be desirable to have, or does not really need to have.

A system design document normally describes on paper through pictures and text what the proposed software program will look like and how it will work.

Aspects to be considered in setting up a computerized database are:

1. Philosophy on which interpretation of results is going to be based:
   a. Critical or standard values
   b. Sufficiency ranges
   c. DRIS system
   and organization of the database to meet the user needs. These will have a tremendous impact on the choice or development of both the data processing system and the accompanying data acquisition forms.
2. Need for historical and on-line tracking of results.
3. Evaluation of computer equipment needs including type(s) of both computers and printers as well as network and modem requirements, if electronic dissemination of data is planned.
4. Actual computer program. Based on the findings of the functionality and attribute definition, the laboratory can proceed with purchasing an existing program or develop its own internally.
5. Design of corresponding forms. This should include both internal (e.g., receiving and laboratory sheets) and external (e.g., information sheet) forms.

The degree to which laboratories should computerize depends on sample volume, location, and services to be offered. Generally, high sample volume laboratories offering many types of analyses have a need for more computer sophistication and automation than low sample volume laboratories. However, often automation in smaller laboratories can lead to realization of many internal efficiencies, in spite of the originally high cost of the computerized system development.

One of the goals in computerizing a system is to eliminate manual tasks including interpretation of results. Often, interpretation of plant analysis data is not straightforward and all possible causes of a physiological phenomenon cannot be described via a mechanized system. Thus, serious consideration must be given to the education of the clientele both in the use of the information sheets, so that the right information is entered for interpretation of the results, and in the method of interpretation of the results and derivation of recommendations.

Most of the laboratories involved in routine plant analysis are using some form of a data processing system. To illustrate how a computerized system is used in a plant analysis laboratory, the Plant analysis module (PAM) in the Enviro-Test Agricultural Services Analytical Resource Management System (ARMS) is discussed.

Figure 27.1 provides a flowchart of the plant testing computer system and associated functions developed by Enviro-Test Laboratories Agricultural Ser-

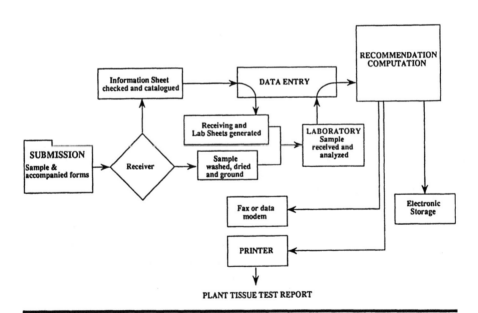

FIGURE 27.1   Plant analysis system at Enviro-Test Laboratories Agricultural Services.

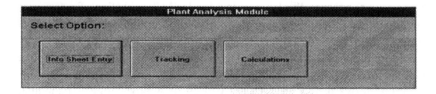

FIGURE 27.2 Computer screen showing the three main modules.

vices in 1995. The system currently consists of three modules: data entry, tracking and calculations (Figure 27.2).

DATA ENTRY/ACQUISITION

Normally, there are two stages in the data acquisition process: (1) entry of all inputs required to calculate fertilizer recommendations and disseminate the information to the user, and (2) entry of all analytical data.

INFORMATION SHEET ENTRY/ACQUISITION

The simplest method of transferring information from an information sheet to a computer is by typing the data on a keyboard. Although simple, errors in transcription may occur and speed of data entry is relatively slow. All inputs used to calculate the corresponding recommendations must be included in an appropriately designed form for the adopted system. An example of an information sheet used by Enviro-Test Laboratories Agricultural Services is shown in Figure 27.3. The information on this sheet is manually transcribed to two screens. The first screen contains the user identification information (Figure 27.4). The program is linked to a client database that allows filling in the user identification information by simply entering a code. The first screen also allows for multiple entries for the same user without the need for re-entering the user identification every time. The second screen contains all the pertinent crop information to arrive at recommendations and is specific to each field (Figure 27.5). Double entry of the same information into the system ensures integrity of information transcription.

A bar code system, such as the one used by PIVOT Laboratories in Australia (Figure 27.6), further automates data acquisition. Although this method is faster, errors in filling the forms can also occur, especially from an unsophisticated user.

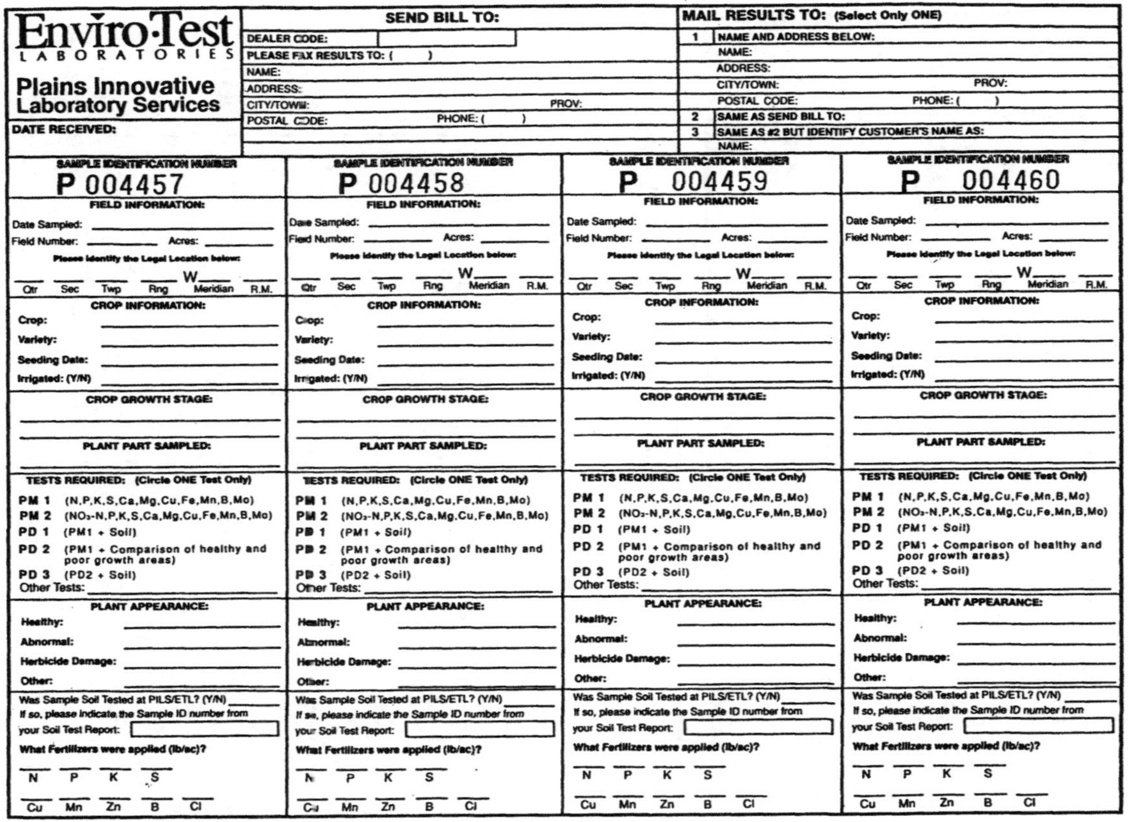
Enviro·Test LABORATORIES
Plains Innovative Laboratory Services
DATE RECEIVED:

SEND BILL TO:
DEALER CODE:
PLEASE FAX RESULTS TO: ( )
NAME:
ADDRESS:
CITY/TOWN: PROV:
POSTAL CODE: PHONE: ( )

MAIL RESULTS TO: (Select Only ONE)
1 NAME AND ADDRESS BELOW:
NAME:
ADDRESS:
CITY/TOWN: PROV:
POSTAL CODE: PHONE: ( )
2 SAME AS SEND BILL TO:
3 SAME AS #2 BUT IDENTIFY CUSTOMER'S NAME AS:
NAME:

SAMPLE IDENTIFICATION NUMBER
P 004457
FIELD INFORMATION:
Date Sampled:
Field Number: Acres:
Please identify the Legal Location below:
Qtr Sec Twp Rng W Meridian R.M.
CROP INFORMATION:
Crop:
Variety:
Seeding Date:
Irrigated: (Y/N)
CROP GROWTH STAGE:
PLANT PART SAMPLED:
TESTS REQUIRED: (Circle ONE Test Only)
PM 1 (N,P,K,S,Ca,Mg,Cu,Fe,Mn,B,Mo)
PM 2 (NO_3-N,P,K,S,Ca,Mg,Cu,Fe,Mn,B,Mo)
PD 1 (PM1 + Soil)
PD 2 (PM1 + Comparison of healthy and poor growth areas)
PD 3 (PD2 + Soil)
Other Tests:
PLANT APPEARANCE:
Healthy:
Abnormal:
Herbicide Damage:
Other:
Was Sample Soil Tested at PILS/ETL? (Y/N)
If so, please indicate the Sample ID number from your Soil Test Report:
What Fertilizers were applied (lb/ac)?
N P K S
Cu Mn Zn B Cl

SAMPLE IDENTIFICATION NUMBER
P 004458
FIELD INFORMATION:
Date Sampled:
Field Number: Acres:
Please identify the Legal Location below:
Qtr Sec Twp Rng W Meridian R.M.
CROP INFORMATION:
Crop:
Variety:
Seeding Date:
Irrigated: (Y/N)
CROP GROWTH STAGE:
PLANT PART SAMPLED:
TESTS REQUIRED: (Circle ONE Test Only)
PM 1 (N,P,K,S,Ca,Mg,Cu,Fe,Mn,B,Mo)
PM 2 (NO_3-N,P,K,S,Ca,Mg,Cu,Fe,Mn,B,Mo)
PD 1 (PM1 + Soil)
PD 2 (PM1 + Comparison of healthy and poor growth areas)
PD 3 (PD2 + Soil)
Other Tests:
PLANT APPEARANCE:
Healthy:
Abnormal:
Herbicide Damage:
Other:
Was Sample Soil Tested at PILS/ETL? (Y/N)
If so, please indicate the Sample ID number from your Soil Test Report:
What Fertilizers were applied (lb/ac)?
N P K S
Cu Mn Zn B Cl

SAMPLE IDENTIFICATION NUMBER
P 004459
FIELD INFORMATION:
Date Sampled:
Field Number: Acres:
Please identify the Legal Location below:
Qtr Sec Twp Rng W Meridian R.M.
CROP INFORMATION:
Crop:
Variety:
Seeding Date:
Irrigated: (Y/N)
CROP GROWTH STAGE:
PLANT PART SAMPLED:
TESTS REQUIRED: (Circle ONE Test Only)
PM 1 (N,P,K,S,Ca,Mg,Cu,Fe,Mn,B,Mo)
PM 2 (NO_3-N,P,K,S,Ca,Mg,Cu,Fe,Mn,B,Mo)
PD 1 (PM1 + Soil)
PD 2 (PM1 + Comparison of healthy and poor growth areas)
PD 3 (PD2 + Soil)
Other Tests:
PLANT APPEARANCE:
Healthy:
Abnormal:
Herbicide Damage:
Other:
Was Sample Soil Tested at PILS/ETL? (Y/N)
If so, please indicate the Sample ID number from your Soil Test Report:
What Fertilizers were applied (lb/ac)?
N P K S
Cu Mn Zn B Cl

SAMPLE IDENTIFICATION NUMBER
P 004460
FIELD INFORMATION:
Date Sampled:
Field Number: Acres:
Please identify the Legal Location below:
Qtr Sec Twp Rng W Meridian R.M.
CROP INFORMATION:
Crop:
Variety:
Seeding Date:
Irrigated: (Y/N)
CROP GROWTH STAGE:
PLANT PART SAMPLED:
TESTS REQUIRED: (Circle ONE Test Only)
PM 1 (N,P,K,S,Ca,Mg,Cu,Fe,Mn,B,Mo)
PM 2 (NO_3-N,P,K,S,Ca,Mg,Cu,Fe,Mn,B,Mo)
PD 1 (PM1 + Soil)
PD 2 (PM1 + Comparison of healthy and poor growth areas)
PD 3 (PD2 + Soil)
Other Tests:
PLANT APPEARANCE:
Healthy:
Abnormal:
Herbicide Damage:
Other:
Was Sample Soil Tested at PILS/ETL? (Y/N)
If so, please indicate the Sample ID number from your Soil Test Report:
What Fertilizers were applied (lb/ac)?
N P K S
Cu Mn Zn B Cl

FIGURE 27.3 Enviro-Test Laboratories Agricultural Services plant analysis information sheet.

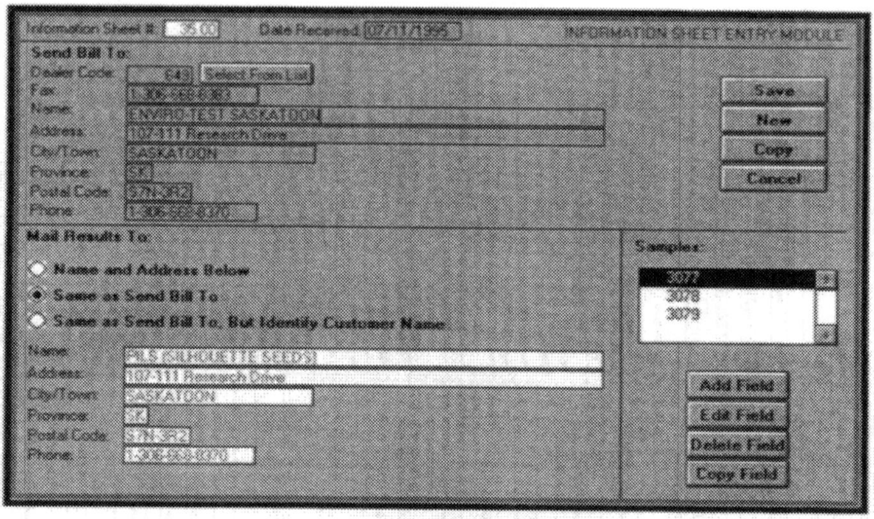

FIGURE 27.4   **Data entry screen for client identification from the plant analysis information sheet.**

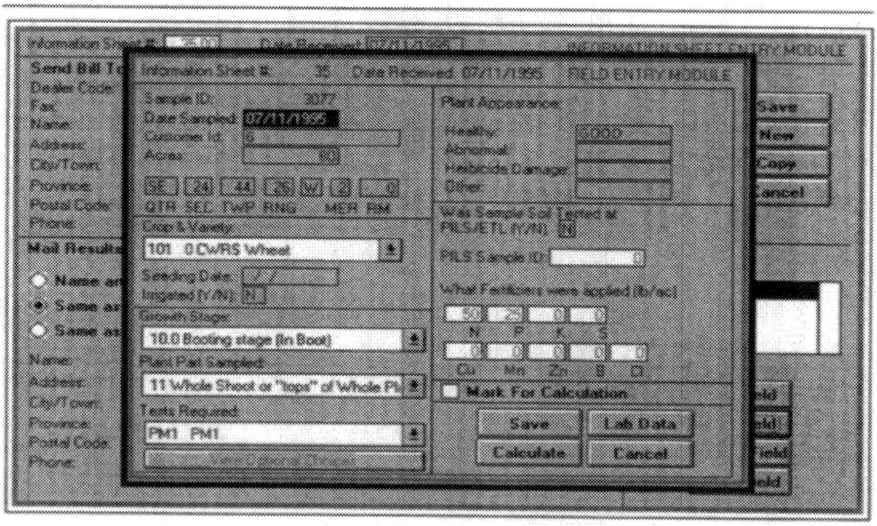

FIGURE 27.5   **Data entry screen for crop information from the plant analysis information sheet.**

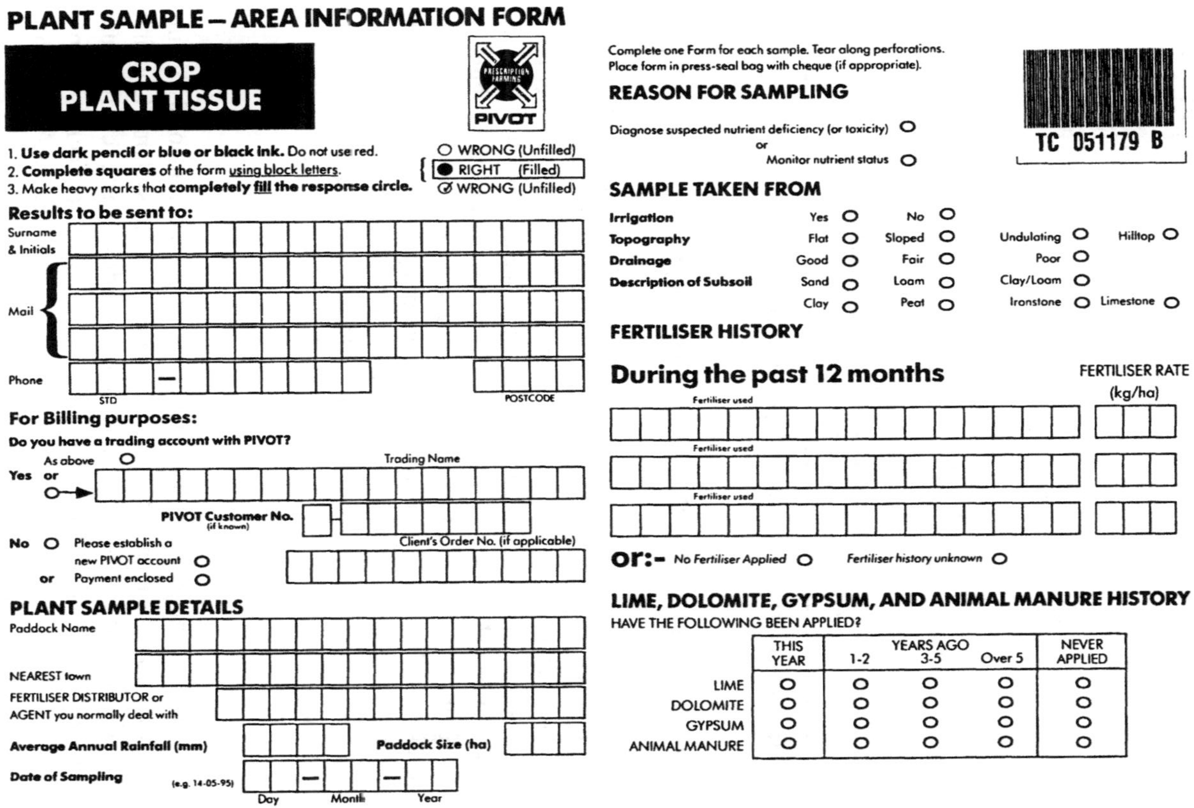

PLANT SAMPLE – AREA INFORMATION FORM

CROP
PLANT TISSUE

PIVOT

1. Use dark pencil or blue or black ink. Do not use red.
2. Complete squares of the form using block letters.
3. Make heavy marks that completely fill the response circle.

O WRONG (Unfilled)
● RIGHT (Filled)
Ø WRONG (Unfilled)

Results to be sent to:
Surname & Initials
Mail
Phone — STD — POSTCODE

For Billing purposes:
Do you have a trading account with PIVOT?
Yes — As above O or O→ Trading Name
PIVOT Customer No. (if known)
No O Please establish a new PIVOT account O
or Payment enclosed O
Client's Order No. (if applicable)

PLANT SAMPLE DETAILS
Paddock Name
NEAREST town
FERTILISER DISTRIBUTOR or AGENT you normally deal with
Average Annual Rainfall (mm)
Paddock Size (ha)
Date of Sampling (e.g. 14-05-95) Day — Month — Year

Complete one Form for each sample. Tear along perforations.
Place form in press-seal bag with cheque (if appropriate).

TC 051179 B

REASON FOR SAMPLING
Diagnose suspected nutrient deficiency (or toxicity) O
or
Monitor nutrient status O

SAMPLE TAKEN FROM
Irrigation: Yes O No O
Topography: Flat O Sloped O Undulating O Hilltop O
Drainage: Good O Fair O Poor O
Description of Subsoil: Sand O Loam O Clay/Loam O Clay O Peat O Ironstone O Limestone O

FERTILISER HISTORY
During the past 12 months
Fertiliser used — FERTILISER RATE (kg/ha)
Fertiliser used
Fertiliser used
or:- No Fertiliser Applied O Fertiliser history unknown O

LIME, DOLOMITE, GYPSUM, AND ANIMAL MANURE HISTORY
HAVE THE FOLLOWING BEEN APPLIED?

	THIS YEAR	YEARS AGO 1-2	3-5	Over 5	NEVER APPLIED
LIME	O	O	O	O	O
DOLOMITE	O	O	O	O	O
GYPSUM	O	O	O	O	O
ANIMAL MANURE	O	O	O	O	O

FIGURE 27.6 Information sheet utilizing bar codes and a scanning device for transcription of information.

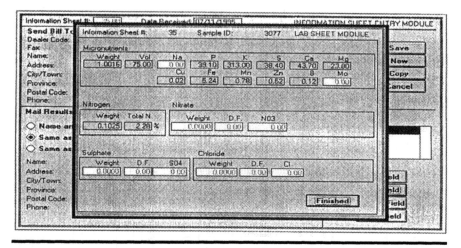

FIGURE 27.7 Data entry screen for analytical results.

ANALYTICAL DATA ENTRY/ACQUISITION

Again, the simplest method of transferring test results from an instrument to a computer is by typing the data on a keyboard. Electronic data capture offers many advantages since it is well suited for automation, eliminates transcription errors, provides immediate data access, and usually enhances productivity and accuracy. However, the investment required must be justified by the volume of samples to be processed by the laboratory. An example of data entry into the plant analysis module used by Enviro-Test Laboratories Agricultural Services is shown in Figure 27.7. In this system the data entry screens are selected by the program based on the analytical package selected by the user.

TRACKING

The progress of the sample through the laboratory can be continuously monitored through the establishment of a tracking module in the system. Tracking is used primarily for client service, although historical data can be retrieved and summarized when necessary. There are a variety of ways to track a sample through the system. The ones chosen by Enviro-Test Laboratories Agricultural Services are shown in Figures 27.8 and 27.9.

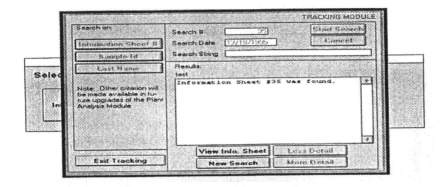

FIGURE 27.8 Computer screen allowing the user to track a plant sample in the system and view the plant analysis report.

CALCULATIONS

This module controls the physical processing of the data in the computer system, from the printing of the lab sheets based on the input of information sheets to processing the lab data into a form that can be printed in the final reports. The module contains all databases required to classify the field information, the parameters required to calculate or determine the correct fertilizer recommendations, and all comments that will be printed in the plant analysis report (Figure 27.10).

The recommendations are based on the critical or standard value concept. Therefore, the database is structured to identify crop, growth stage, and plant part sampled. Crops with the same growth stages are classified in the same growth stage group although the criteria and recommendations may be different for different crops. If a plant sample is submitted for analysis from a crop that is not in a diagnostic growth stage, a statement appears in the comments section of the plant analysis report to that effect. Further, if a crop is at a stage when an identified deficiency cannot be corrected, the client is also informed and is advised to follow up with a soil test.

For example, the growth stages of all cereals can be described using either Feeke's or Zadock's scales of growth stages. In the Enviro-Test Laboratories Agricultural Services system, all cereals have been assigned to Growth Stage Group 1 (Figure 27.11). A partial list of how growth stages for Growth Stage Group 1 are structured in the system is shown in Figure 27.12.

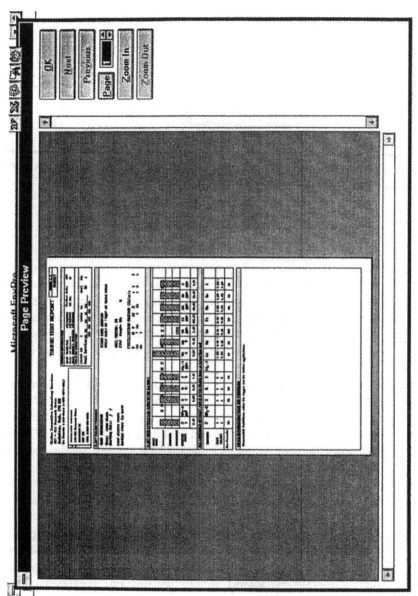

FIGURE 27.9   Computer screen allowing the user to view a plant analysis report.

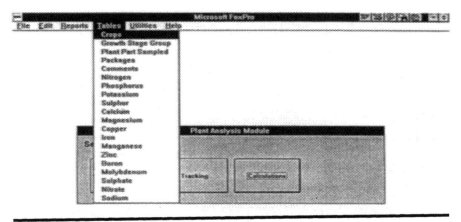

FIGURE 27.10 Databases for the determination of fertilizer recommendations.

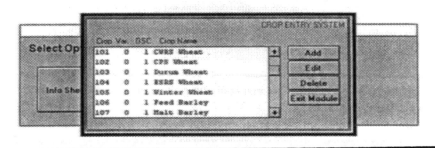

FIGURE 27.11 Computer screen containing all crops in the system. Each crop is assigned a code and a growth stage group. Different varieties of the same crop can be treated separately.

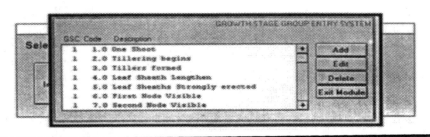

FIGURE 27.12 Computer screen containing growth stages for each growth stage group.

Plant Analysis Module			
Plant Abbrev.	PPS Code	Description	↑
YEB+2	14	the next older leaf blade below YEB+1	
YFL	15	Youngest folded leaf blade	
YMB	16	Youngest mature leaf blade	
YMB+1	17	the next older leaf blade below YMB	
YMB+2	18	the next older leaf blade below YMB+1	
YMHL	19	Youngest mature heart leaf	
YOL	20	Youngest open leaf blade below YOL	

FIGURE 27.13 Computer screen containing all plant parts for which diagnostic criteria are included in the system.

Plant Analysis Module		
Comment Code	Comment	↑
1	Application of Nitrogen fertilizer at this growth stage will not contribute to yield increase but can en	
2	Application of Nitrogen fertilizer at this growth stage is not recommended. A soil test is recommen	
3	There are no Nitrogen diagnostic criteria for the growth stage and plant part sampled.	
4	Phosphorus applications after seeding are generally inefficient. A soil test is recommened to estab	
5	There are no Phosphorus diagnostic criteria for the growth stage and plant part sampled.	
6	Potassium applications after seeding are generally inefficient. A soil test is recommended to estab	
7	There are no Potassium diagnostic criteria for the growth stage and plant part sampled.	

FIGURE 27.14 Computer screen containing all comments that are generated by the system.

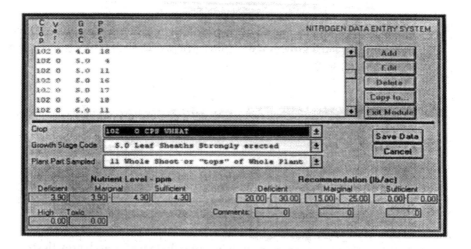

FIGURE 27.15 Computer screen displaying an example of the database structure in the system.

The plant parts abbreviations and descriptions in the system are shown in Figure 27.13.

A plant analysis data processing system by its nature contains a series of explanatory comments of the conditions under which criteria have been derived or whether fertilizer should be applied, type(s) of fertilizers to be used, why a certain nutrient may not be at the optimum range, etc. (Figure 27.14).

The database itself is structured to fulfill the requirements dictated by the choice of the philosophy for interpreting the results. To maintain integrity of the database, the module containing all agronomic data can either be protected by a password or series of passwords or in more sophisticated systems by the choice of the system server. In any case, documentation of any additions, modifications, or deletions from the agronomic database is essential in maintaining the integrity of the data processing system. An example of a database table from the Enviro-Test Laboratories Agricultural Services system is illustrated in Figure 27.15. Only the laboratory agronomist has access to this module and all changes are automatically tracked by the computer.

PLANT AND RELATED REFERENCE MATERIALS FOR DATA QUALITY CONTROL OF ELEMENTAL CONTENT

I

Milan Ihnat

INTRODUCTION

This appendix is complementary and integral to Chapter 26 on Reference Materials for Data Quality Control. It is intended to facilitate location and selection of appropriate plant Reference Materials for utilization for analytical data quality control as detailed in Chapter 26. Presented here, in tabular format, is a fairly comprehensive, up-to-date, listing of natural matrix plant and related Reference Materials, generally from major government agency producers, with quoted concentration values for native elemental content. As mentioned in Chapter 26, while extractable or bioavailable elemental concentrations are frequently of interest, only the control of total elemental concentrations is addressed because of the usual requirement in plant analysis for total contents and the lack of Reference Materials for extractable analytes. Table 1 lists the names and addresses of some of the major (mainly government) producers and suppliers producing and distributing plant and related plant-product Reference Materials for chemical composition quality control in plant analysis. In Table 2 are presented plant and related Reference Materials for chemical composition data quality control available from government agency and other selected pro-

1-57444-124-8/98/$0.00+$.50
© 1998 by CRC Press LLC

ducers/suppliers, including producer and material codes and quoted elements. The materials are listed in Table 3 alphabetically by element and, for each element, in increasing order of concentration to assist the analyst to locate a material of appropriate natural matrix composition and element concentration level. Finally, the appendix concludes (Tables 4 and 5) with an example of element concentration values and analytical methods used in characterization of one Reference Material (Reference Material Corn Bran, NIST RM 8433) from a typical report of investigation.

The Reference Materials listed here are generally applicable to all analytical methods and elements reported in this manual insofar as the analytes are represented by existing Reference Materials. For completeness of information, all elements quoted for the materials are listed here, encompassing a number of elements not usually of interest to or addressed by plant researchers, but obviously measured by others. Thus, included are elements other than those dealt with in this manual, for future use as research and method development is extended to elements other than those of current interest to plant scientists. Knowledge of the availability of control materials for less common elements may spur interest in such elements in response to environmental, nutritional, and food and feed safety concerns as well as comprehensive elemental characterization of crops.

As reliability and absolute confidence in the stated characteristics of Reference Materials is a basic critical criterion for their use for quality control, solely *bona fide* Reference Materials are listed. No presentation has been made of in-house, local, regional or other "uncertified" reference products, even though such well-prepared and characterized homogeneous materials can be of good quality and useful for day-to-day or more frequent monitoring of precision. Certified, *bona fide* Reference Materials are reserved for monitoring analytical bias and the products listed here are generally believed to be good choices for routine quality control.

Arising from different agencies following different preparation and certification approaches, all products denoted Reference Materials or Certified Reference Materials are not to be expected to adhere to uniform criteria of quality. Concentration values may be denoted, among many other terms, certified, recommended, or informational. In this appendix, elements listed are those for which the issuing organization or the producer indicates to have certified or recommended concentration values. Other elements with concentration values listed by the producers in certificates or reports of analysis as informational or otherwise are not included here.

An extremely wide range of concentration values is available, for a large number of agronomically, nutritionally, toxicologically, and environmentally important elements. Recent developments (Ihnat, 1994) have augmented the world repertoire of Reference Materials and have contributed substantially to the elemental database with concentration information for, among others, agronomically pertinent elements such as aluminum (Al), boron (B), barium (Ba), iodine (I), molybdenum (Mo), nitrogen (N), sulfur (S), and vanadium (V). For example, best-estimate values are now available for N in plant products ranging from 0.02 to 14.68%, making the current collection of Reference Materials valuable for calibration of nitrogen analyzers and for confirming analytical methods used in analysis of agricultural and food commodities for this nutritionally and economically important element. Products, such as Corn Bran NIST RM 8433), Corn Starch (NIST RM 8432), and Microcrystalline Cellulose (NIST RM 8416), contain very low elemental concentrations and could conceivably serve as real sample blanks in some analytical procedures.

With respect to the availability of plant and other biological and environmental Reference Materials from government agencies, the extensive compilations of Muramatsu and Parr (1985), Cortes Toro et al. (1990), International Atomic Energy Agency (1995) and Cantillo (1995) are very useful sources of information. Together, these sources of information provide a rather complete, up-to-date listing of Reference Materials, including descriptions and concentration data as well as ordering information. Individual catalogues and reports of issuing agencies [e.g., Bowman (1994), International Atomic Energy Agency (1994), Trahey (1995), and European Commission (1996) as well as information from commercial suppliers should also be consulted.

Information in the tables is based on data from the compilations of Bowen (1985), Muramatsu and Parr (1985), Ihnat (1988), Cortes Toro et al. (1990), International Atomic Energy Agency (1994, 1995), Ihnat (1994), Cantillo (1995), Trahey (1995), European Commission (1996) as well as from individual certificates and reports of analysis issued by the producers.

The information in the tables in this appendix is provided for the convenience of the analyst and should be used only as guides for Reference Material selection and use. Although an effort has been made to ensure that this information is reliable, the analytical data herein must not be used for data quality control.

Adherence should be made to the philosophy that the latest, appropriate certificates or reports of analysis or other relevant documents and publications accompanying or pertaining to the Reference Material be consulted for concentration values, uncertainties, correct usage of the material, and interpretation of analytical results.

TABLE 1 Producers and suppliers of reference materials for elemental composition quality control in plant analysis[a]

Code	Source
AAFC	Dr. M. Ihnat, Eastern Cereal and Oilseed Research Centre, Agriculture and Agri-Food Canada, Ottawa, Ontario K1A 0C6, Canada (see NIST).
AMM	Faculty of Physics and Nuclear Techniques, University of Mining and Metallurgy, Al Mickiewicza 30, 30-059 Krakow, Poland.
ARC	Food Research Institute, Laboratory of Food Chemistry, Agricultural Research Centre of Finland, SF-31600 Jokioinen, Finland.
BCR	Institute for Reference Materials and Measurements (IRMM), Retieseweg, B-2440 Geel, Belgium.
BOWEN	Dr. H.J.M. Bowen, West Down, West Street, Winterborne Kingston, Dorset DT11 9AT, Great Britain.
CALNRI	Central Analytical Laboratory, Nuclear Research Institute Rez plc, 250 68 Rez, Czech Republic.
CANMET	Canadian Certified Reference Materials Project, Canada Centre for Mineral and Energy Technology, Natural Resources Canada, 555 Booth Street, Ottawa, Ontario K1A 0G1, Canada.
CSRM	Pb-Anal, Garbiarska 2, 040 01 Kosice, Slovakia.
DL	AG Dillinger Hüttenwerke, Postfach 1580, D-66748 Dilingen-Saar, Germany.
GBW	Office of CRMs, National Research Centre for Certified Reference Materials, No. 18 Bei San Huan Dong Hu, Hepingjie, 100013 Beijing, China.
IAEA	Analytical Quality Control Services, International Atomic Energy Agency, P.O. Box 100, A-1400 Wien, Austria.
ICHTJ	Commission of Trace Analysis of the Committee for Analytical Chemistry of the Polish Academy of Sciences, Department of Analytical Chemistry, Institute of Nuclear Chemistry and Technology, ul. Dorodna 16, 03-195 Warszawa, Poland.
LIVSVER	Chemistry Division 2, Swedish National Food Administration, P.O. Box 622, S-751 26 Uppsala, Sweden.
NIES	Division of Environmental Chemistry, National Institute for Environmental Studies, 16-2 Onagawa, Tsukuba, Ibaraki 305, Japan.

TABLE 1 Producers and suppliers of reference materials for elemental composition quality control in plant analysis[a] (continued)

Code	Source
NIST	Standard Reference Materials Program, National Institute of Standards and Technology, Room 204, Building 202, Gaithersburg, MD 20899.
WSPTP	Western States Proficiency Testing Program. Dr. R.O. Miller, Department of Land, Air, and Water Resources, University of California, Davis, CA 95616-8627.

[a] Primarily major government agencies with inclusion of some academic, commercial, and private sources. Reference Materials are also available from other distributors, which may be more convenient sources for purchase.

TABLE 2 Plant and related reference materials for chemical composition available from, mainly, government agency suppliers[a]

Material and source[b]	Code[b]	Quoted elements[c]
AMM		
Cabbage leaves	AMM-CL-1	B, Ca, Cd, Cu, Fe, K, Mg, Mn, Na, Ni, Pb, Rb, Se, Sr, Zn
ARC		
Carrot powder	ARC/CL-CP	B, Ca, Cd, Fe, K, Mg, Mn, Mo, Na, Zn
Potato powder	ARC/CL-PP	Ca, Cd, Cu, Fe, Mg, Mn, Mo, Ni, Pb, Zn
Wheat flour	ARC/CL-WF	Ca, Cd, Cr, Cu, Fe, K, Mg, Mn, Mo, Ni, P, Pb, Se, Zn
BCR		
Aquatic plant (*L. major*)	BCR-CRM-060	Al, Cd, Cu, Hg, Mn, Pb, Zn
Aquatic plant (*P. riparioides*)	BCR-CRM-061	Al, Cd, Cu, Hg, Mn, Pb, Zn
Olive leaves (*O. europaea*)	BCR-CRM-062	Al, Cd, Cu, Hg, Mn, Pb, Zn
Beech leaves	BCR-CRM-100	Al, Ca, Cl, K, Mg, N, P, S
Spruce needles	BCR-CRM-101	Al, Ca, Cl, Mg, Mn, N, P, S, Zn

TABLE 2   Plant and related reference materials for chemical composition available from, mainly, government agency suppliers[a] (continued)

Material and source[b]	*Code[b]*	*Quoted elements[c]*
BCR (continued)		
Hay powder	BCR-CRM-129	Ca, I, K, Mg, N, P, S, Zn
Whole meal flour	BCR-CRM-189	Cd, Cu, Fe, Mn, Pb, Se, Zn
Whole rapeseed (medium level)	BCR-CRM-190	S
Brown bread	BCR-CRM-191	Cd, Cu, Fe, Mn, Pb, Zn
Single cell protein	BCR-CRM-273	Ca, Fe, K, Mg, P
Single cell protein	BCR-CRM-274	As, Cd, Co, Cu, Mn, Pb, Se, Zn
Sea lettuce (*Ulva lactuca*)	BCR-CRM-279	As, Cd, Cu, Pb, Se, Zn
Rye grass	BCR-CRM-281	As, B, Cd, Cu, Hg, Mn, Mo, Ni, Pb, Sb, Se, Zn
Whole rapeseed (low level)	BCR-CRM-366	S
Whole rapeseed (high level)	BCR-CRM-367	S
Rye flour	BCR-CRM-381	Ca, Cl, K, Mg, N, Na
Wheat flour	BCR-CRM-382	N
Haricots verts (beans)	BCR-CRM-383	Ca, K, N, Na
White clover	BCR-CRM-402	As, Co, Mo, Se
Plankton	BCR-CRM-414	As, Cd, Cr, Cu, Hg, Mn, Ni, Pb, Se, V, Zn
Lichen	BCR-CRM-482	Al, Cd, Cr, Cu, Hg, Ni, Pb, Zn
Aquatic plant (*Trapa natans*)	BCR-CRM-596	Cr
BOWEN		
Kale	BOWEN'S KALE	B, Br, Ca, Cl, Co, Cs, Cu, F, Fe, Hg, K, La, Mg, Mo, N, Na, P, Rb, S, V, Zn
CALNRI		
Yeast (Torulopsis E.)	CALNRI-TY-1	As, Cd, Co, Cu, Fe, Hg, K, Mg, Mn, Na, Ni, P, Rb, Sb, Zn
Yeast (Torulopsis E.)	CALNRI-TY-2	Cd, Co, Cr, Cu, Fe, Hg, K, Mg, Mn, Na, Ni, P, Rb, Sb, Zn
Yeast (Torulopsis E.)	CALNRI-TY-3	Co, Cr, Cu, Fe, Hg, K, Mg, Mn, Na, Ni, P, Rb, Sb, Zn

TABLE 2 Plant and related reference materials for chemical composition available from, mainly, government agency suppliers[a] (continued)

Material and source[b]	Code[b]	Quoted elements[c]
CALNRI (continued)		
Yeast (Torulopsis E.)	CALNRI-TY-4	Cd, Co, Cr, Cu, Fe, Hg, K, Mn, Na, P, Rb, Sb, Zn
CANMET		
Spruce twigs and needles	CANMET-CLV-1	U
Spruce twigs and needles	CANMET-CLV-2	U
CSRM		
Green algae (*C. kessleri*)	CSRM-12-2-02	Ca, Cd, Co, Cr, Cu, Fe, K, Mg, Mn, Na, P, Pb, Zn
Lucerne P-Alfalfa	CSRM-12-2-03	Ba, Ca, Cd, Cu, Fe, K, Mg, Mn, Na, Ni, P, Pb, Sr, Zn
Wheat bread flour	CSRM-12-2-04	Ca, Cd, Cu, Fe, K, Mg, Mn, P, Pb, Zn
Rye bread flour	CSRM-12-2-05	Ca, Cd, Cu, Fe, K, Mg, Mn, P, Pb, Sr, Zn
DL		
Rice straw ash	DL-5702	Al, Ca, Fe, K, Mg, Mn, Na, P, Si, Ti
GBW		
Bush branches and leaves	GBW-07602	Ag, Al, As, B, Ba, Be, Br, Ca, Cd, Ce, Co, Cr, Cs, Cu, Eu, F, Fe, Hf, K, La, Li, Mg, Mn, Mo, N, Na, Ni, P, Pb, Rb, S, Sb, Sc, Se, Si, Sm, Sr, Th, Ti, V, Yb, Zn
Bush branches and leaves	GBW-07603	Ag, Al, As, B, Ba, Be, Bi, Br, Ca, Ce, Co, Cr, Cs, Cu, Eu, F, Fe, K, La, Li, Mg, Mn, Mo, N, Na, Nd, Ni, P, Pb, Rb, S, Sb, Sc, Se, Si, Sm, Sr, Tb, Th, Ti, V, Y, Yb, Zn

TABLE 2   Plant and related reference materials for chemical composition available from, mainly, government agency suppliers[a] (continued)

Material and source[b]	Code[b]	Quoted elements[c]
GBW (continued)		
Poplar leaves	GBW-07604	Al, As, B, Ba, Be, Bi, Br, Ca, Cd, Ce, Co, Cr, Cs, Cu, Eu, F, Fe, Hg, K, La, Li, Mg, Mn, Mo, N, Na, Ni, P, Pb, Rb, S, Sb, Sc, Se, Si, Sm, Sr, Th, Ti, Y, Yb, Zn
Tea	GBW-07605	As, B, Ba, Be, Bi, Br, Ca, Cd, Ce, Co, Cr, Cs, Cu, Eu, F, Fe, K, La, Mg, Mn, Mo, N, Na, Ni, P, Pb, Rb, S, Sb, Sc, Sm, Sr, Th, Ti, Y, Yb, Zn
Peach leaves	GBW-08501	As, B, Ba, Cd, Cr, Cu, Fe, Hg, K, Mg, Mn, Pb, Sr, Zn
Rice flour	GBW-08502	As, Ca, Cd, Cu, Fe, K, Mg, Mn, Na, Se, Zn
Wheat flour	GBW-08503	As, Ca, Cd, Cu, Fe, K, Mg, Mn, Pb, Zn
Cabbage	GBW-08504	As, Ca, Cd, Cu, Fe, K, Mg, Mn, N, Na, P, Pb, Se, Sr, Zn
Tea	GBW-08505	As, Ba, Ca, Cd, Ce, Cu, Fe, K, La, Mg, Mn, N, Na, Ni, P, Pb, Rb, S, Sb, Se, Sr, Th, Zn
Fluoride composition in corn	GBW-08506	F
Fluoride composition in corn	GBW-08507	F
Mercury in rice	GBW-08508	Cu, Fe, Hg, Mn, Zn
Codonopsis P.	GBW-09501	Ba, Ca, Cu, Fe, Mn, Sr, Zn
IAEA		
Lichen	IAEA-336	As, Br, Cd, Ce, Co, Cs, Cu, Fe, Hg, K, Mn, Rb, Sb, Sc, Se, Sm, Zn
Grass	IAEA-373	Th

TABLE 2 Plant and related reference materials for chemical composition available from, mainly, government agency suppliers[a] (continued)

Material and source[b]	Code[b]	Quoted elements[c]
IAEA (continued)		
Hay powder	IAEA-V-10	Ba, Br, Ca, Cd, Co, Cr, Cu, Fe, Hg, Mg, Mo, Ni, P, Pb, Rb, Sc, Sr, Zn
Rye flour	IAEA-V-8	Br, Ca, Cl, Cu, Fe, K, Mg, Mn, P, Rb, Zn
Cotton cellulose	IAEA-V-9	Ba, Ca, Cl, Cr, Cu, Hg, Mg, Mn, Mo, Na, Ni, Pb, Sr
ICHTJ		
Oriental tobacco leaves	ICHTJ-CTA-OTL-1	Al, As, Ba, Br, Ca, Cd, Ce, Co, Cr, Ce, Cu, Eu, K, La, Li, Mg, Mn, Ni, P, Pb, Rb, S, Se, Sm, Sr, Tb, Th, V, Zn
LIVSVER		
Cantharellus T. (fungus)	LIVSVER-FUNGUS	Cd, Co, Cr, Cu, Fe, Mn, Ni, Pb, Zn
NIES		
Pepperbush	NIES-CRM-1	As, Ba, Ca, Cd, Co, Cu, Fe, K, Mg, Mn, Na, Ni, Pb, Rb, Sr, Zn
Rice flour	NIES-CRM-10A	Ca, Cd, Cu, Fe, K, Mg, Mn, Mo, Na, Ni, P, Rb, Zn
Rice flour	NIES-CRM-10B	Ca, Cd, Cu, Fe, K, Mg, Mn, Mo, Na, Ni, P, Rb, Zn
Rice flour	NIES-CRM-10C	Ca, Cd, Cu, Fe, K, Mg, Mn, Mo, Na, Ni, P, Rb, Zn
Chlorella	NIES-CRM-3	Ca Co, Cu, Fe, K, Mg, Mn, Sr, Zn
Tea leaves	NIES-CRM-7	Al, Ca, Cd, Cu, K, Mg, Mn, Na, Ni, Pb, Zn
Sargasso	NIES-CRM-9	Ag, As, Ca, Cd, Co, Cu, Fe, K, Mg, Mn, Na, Pb, Rb, Sr, V, Zn

TABLE 2 Plant and related reference materials for chemical composition available from, mainly, government agency suppliers[a] (continued)

Material and source[b]	Code[b]	Quoted elements[c]
NIST		
Corn stalk (*Zea mays*)	NIST-RM-8412	Ca, Cl, Cu, Fe, K, Mg, Mn, Na, Se, Sr, Zn
Corn kernel (*Zea mays*)	NIST-RM-8413	Ca, Cu, Fe, K, Mg, Mn, Se, Zn
Microcrystalline cellulose	NIST-RM-8416	Al, Cd, Cl, Co, Cu, Mo, N, Ni, Pb, Se
Wheat gluten	NIST-RM-8418	Al, Ba, Ca, Cd, Cl, Co, Cr, Cu, Fe, Hg, I, K, Mg, Mn, Mo, N, Na, Ni, P, Pb, S, Se, Sr, Zn
Corn starch	NIST-RM-8432	Al, Ca, Cd, Cl, Co, Cu, Hg, K, Mg, Mn, Mo, N, Na, Ni, P, Pb, Se, Sr, Zn
Corn bran	NIST-RM-8433	Al, As, B, Ba, Br, Ca, Cd, Cl, Co, Cr, Cu, Fe, Hg, I, K, Mg, Mn, Mo, N, Na, Ni, P, Pb, Rb, S, Se, Sr, V, Zn
Durum wheat flour	NIST-RM-8436	Al, Ba, Br, Ca, Cd, Cl, Co, Cr, Cu, Fe, Hg, I, K, Mg, Mn, Mo, N, Na, Ni, P, Pb, Rb, S, Se, Sr, V, Zn
Hard red spring wheat flour	NIST-RM-8437	Al, Ca, Cl, Cr, Cu, Fe, K, Mg, Mn, Mo, N, Na, P, S, Se, V, Zn
Soft winter wheat flour	NIST-RM-8438	Al, Ca, Cl, Cr, Cu, Fe, K, Mg, Mn, Mo, N, Na, P, S, Se, Zn
Apple leaves	NIST-SRM-1515	Al, As, B, Ba, Ca, Cd, Cl, Cu, Hg, K, Mg, Mn, Mo, N, Na, Ni, P, Pb, Rb, Se, Sr, V, Zn
Peach leaves	NIST-SRM-1547	Al, As, B, Ba, Ca, Cl, Cu, Fe, Hg, K, Mg, Mn, Mo, N, Na, Ni, Pb, Rb, Se, Sr, V, Zn

TABLE 2 Plant and related reference materials for chemical composition available from, mainly, government agency suppliers[a] (continued)

Material and source[b]	Code[b]	Quoted elements[c]
NIST (continued)		
Wheat flour	NIST-SRM-1567	Ca, Cd, Cu, Fe, Hg, K, Mn, Na, Se, Zn
Wheat flour	NIST-SRM-1567a	Al, Ca, Cd, Cu, Fe, K, Mg, Mn, Mo, Na, P, Rb, S, Se, Zn
Rice flour	NIST-SRM-1568	As, Ca, Cd, Co, Cu, Fe, Hg, K, Mn, Na, Rb, Se, Zn
Rice flour	NIST-SRM-1568a	Al, As, Ca, Cd, Cu, Fe, Hg, K, Mg, Mn, Mo, Na, P, Rb, S, Se, Zn
Brewer's yeast	NIST-SRM-1569	Cr
Trace elements in spinach	NIST-SRM-1570	Al, As, Ca, Cr, Cu, Fe, Hg, K, Mn, P, Pb, Rb, Sr, Th, U, Zn
Spinach leaves	NIST-SRM-1570a	Al, As, B, Ca, Cd, Co, Cu, Hg, K, Mn, N, Na, Ni, P, Se, Sr, Th, V, Zn
Citrus leaves	NIST-SRM-1572	Al, As, Ba, Ca, Cd, Cr, Cu, Fe, Hg, I, K, Mg, Mn, Mo, Na, Ni, P, Pb, Rb, S, Sr, Zn
Tomato leaves	NIST-SRM-1573	As, Ca, Cr, Cu, Fe, K, Mn, P, Pb, Rb, Sr, Th, U, Zn
Tomato leaves	NIST-SRM-1573a	Al, As, B, Ca, Cd, Co, Cr, Cu, Fe, Hg, K, Mn, N, Na, Ni, P, Rb, Sb, Se, V, Zn
Pine needles	NIST-SRM-1575	Al, As, Ca, Cr, Cu, Fe, Hg, K, Mn, P, Pb, Rb, Sr, Th, U
Fluoride in vegetation, high	NIST-SRM-2695 H	F
Fluoride in vegetation, low	NIST-SRM-2695 L	F

[a] Based on information in the references listed at the end of the appendix. Refer to International Atomic Energy Agency (1995) and Cantillo (1995) for the most recent, excellent, detailed coverage of plant, and other biological and environmental materials. Former compilations are available by Cortes Toro et al. (1990), Muramatsu and Parr (1985), and Ihnat (1988).

TABLE 2   Plant and related reference materials for chemical composition available from, mainly, government agency suppliers[a] (concluded)

Material and source[b]	*Code[b]*	*Quoted elements[c]*

[b] Source codes are defined in Table 1. Material codes are a combination of producer codes and product identities assigned by the producer. The majority of these Reference Materials are available from the listed issuing organizations. Several older materials may not be available, however, from primary sources, but are included for completeness and for their usefulness and because they may still be available in the secondary market (e.g., from laboratory stocks of obliging colleagues). The user should consider storage, stability, and certificate expiration dates prior to using such materials.

[c] Elements listed are those for which the issuing organization or the producer indicates to have certified or recommended concentration values. Other elements in certificates or reports of analysis with concentration values listed by the producer as informational or otherwise are not included here. Such other concentration values could, however, be of interest and use to the analyst and the original literature should be consulted for details.

TABLE 3   Certified and recommended total elemental concentration values reported for plant and related reference materials listed in order of elements and in increasing order of concentration[a]

Material[b]	*Code[b]*	*Concentration, mg/kg[c]*
Ag: Silver		
Bush branches and leaves	GBW-07602	0.027
Bush branches and leaves	GBW-07603	0.049
Sargasso	NIES-CRM-9	0.31
Al: Aluminum		
Corn bran	NIST-RM-8433	1.01
Corn starch	NIST-RM-8432	1.9
Hard red spring wheat flour	NIST-RM-8437	2.1
Soft winter wheat flour	NIST-RM-8438	2.3
Microcrystalline cellulose	NIST-RM-8416	3.7
Rice flour	NIST-SRM-1568A	4.4
Wheat flour	NIST-SRM-1567A	5.7
Wheat gluten	NIST-RM-8418	10.8
Durum wheat flour	NIST-RM-8436	11.7
Citrus leaves	NIST-SRM-1572	92
Spruce needles	BCR-CRM-101	173
Peach leaves	NIST-SRM-1547	249
Apple leaves	NIST-SRM-1515	286
Spinach leaves	NIST-SRM-1570A	310

TABLE 3 Certified and recommended total elemental concentration values reported for plant and related reference materials listed in order of elements and in increasing order of concentration[a] (continued)

Material[b]	Code[b]	Concentration, mg/kg[c]
Al: Aluminum (continued)		
Beech leaves	BCR-CRM-100	435
Olive leaves (*O. europaea*)	BCR-CRM-062	450
Pine needles	NIST-SRM-1575	545
Tomato leaves	NIST-SRM-1573A	598
Tea leaves	NIES-CRM-7	775
Trace elements in spinach	NIST-SRM-1570	870
Poplar leaves	GBW-07604	1,040
Lichen	BCR-CRM-482	1,103
Oriental tobacco leaves	ICHTJ-CTA-OTL-1	1,740
Bush branches and leaves	GBW-07603	2,000
Bush branches and leaves	GBW-07602	2,140
Rice straw ash	DL-5702	3,234
Aquatic plant (*L. major*)	BCR-CRM-060	4,180
Aquatic plant (*P. riparioides*)	BCR-CRM-061	10,740
As: Arsenic		
Corn bran	NIST-RM-8433	0.002
Apple leaves	NIST-SRM-1515	0.038
Rice flour	GBW-08502	0.051
Cabbage	GBW-08504	0.056
Rye grass	BCR-CRM-281	0.057
Peach leaves	NIST-SRM-1547	0.060
Spinach leaves	NIST-SRM-1570A	0.068
White clover	BCR-CRM-402	0.093
Yeast (Torulopsis E.)	CALNRl-TY-1	0.103
Tomato leaves	NIST-SRM-1573A	0.112
Single cell protein	BCR-CRM-274	0.132
Trace elements in spinach	NIST-SRM-1570	0.150
Tea	GBW-08505	0.191
Pine needles	NIST-SRM-1575	0.21
Wheat flour	GBW-08503	0.22
Tomato leaves	NIST-SRM-1573	0.27
Tea	GBW-07605	0.28
Rice flour	NIST-SRM-1568A	0.29
Peach leaves	GBW-08501	0.34
Poplar leaves	GBW-07604	0.37
Rice flour	NIST-SRM-1568	0.41
Oriental tobacco leaves	ICHTJ-CTA-OTL-1	0.539

TABLE 3   Certified and recommended total elemental concentration values reported for plant and related reference materials listed in order of elements and in increasing order of concentration[a] (continued)

Material[b]	Code[b]	Concentration, mg/kg[c]
As: Arsenic (continued)		
Lichen	IAEA-336	0.639
Bush branches and leaves	GBW-07602	0.95
Bush branches and leaves	GBW-07603	1.25
Pepperbush	NIES-CRM-1	2.3
Sea lettuce (*Ulva lactuca*)	BCR-CRM-279	3.09
Citrus leaves	NIST-SRM-1572	3.1
Plankton	BCR-CRM-414	6.82
Sargasso	NIES-CRM-9	115
B: Boron		
Corn bran	NIST-RM-8433	2.8
Rye grass	BCR-CRM-281	5.9
Carrot powder	ARC/CL-CP	10.1
Tea	GBW-07605	15
Cabbage leaves	AMM-CL-1	20.6
Apple leaves	NIST-SRM-1515	27
Peach leaves	NIST-SRM-1547	29
Tomato leaves	NIST-SRM-1573A	33.3
Bush branches and leaves	GBW-07602	34
Spinach leaves	NIST-SRM-1570A	37.6
Bush branches and leaves	GBW-07603	38
Kale	BOWEN'S KALE	49
Poplar leaves	GBW-07604	53
Ba: Barium		
Wheat gluten	NIST-RM-8418	1.53
Durum wheat flour	NIST-RM-8436	2.11
Corn bran	NIST-RM-8433	2.4
Hay powder	IAEA-V-10	6
Cotton cellulose	IAEA-V-9	9
Tea	GBW-08505	15.7
Bush branches and leaves	GBW-07603	18
Peach leaves	GBW-08501	18.4
Bush branches and leaves	GBW-07602	19
Citrus leaves	NIST-SRM-1572	21
Lucerne P-alfalfa	CSRM-12-2-03	23.4
Poplar leaves	GBW-07604	26
Apple leaves	' NIST-SRM-1515	49

TABLE 3 Certified and recommended total elemental concentration values reported for plant and related reference materials listed in order of elements and in increasing order of concentration[a] (continued)

Material[b]	Code[b]	Concentration, mg/kg[c]
Ba: Barium (continued)		
Tea	GBW-07605	58
Oriental tobacco leaves	ICHTJ-CTA-OTL-1	84.2
Codonopsis P.	GBW-09501	109
Peach leaves	NIST-SRM-1547	124
Pepperbush	NIES-CRM-1	165
Be: Beryllium		
Poplar leaves	GBW-07604	0.021
Tea	GBW-07605	0.034
Bush branches and leaves	GBW-07603	0.051
Bush branches and leaves	GBW-07602	0.056
Bi: Bismuth		
Bush branches and leaves	GBW-07603	0.023
Poplar leaves	GBW-07604	0.027
Tea	GBW-07605	0.063
Br: Bromine		
Rye flour	IAEA-V-8	0.38
Corn bran	NIST-RM-8433	2.3
Bush branches and leaves	GBW-07602	2.4
Bush branches and leaves	GBW-07603	3
Tea	GBW-07605	3.4
Durum wheat flour	NIST-RM-8436	6.6
Poplar leaves	GBW-07604	7.2
Hay powder	IAEA-V-10	8
Oriental tobacco leaves	ICHTJ-CTA-OTL-1	9.28
Lichen	IAEA-336	12.9
Kale	BOWEN'S KALE	24.9
Ca: Calcium		
Corn kernel (*Zea mays*)	NIST-RM-8413	42
Rice flour	GBW-08502	55
Corn starch	NIST-RM-8432	56
Rice flour	NIES-CRM-10B	78
Potato powder	ARC/CL-PP	90
Rice flour	NIES-CRM-10A	93
Rice flour	NIES-CRM-10C	95
Rice flour	NIST-SRM-1568A	118

TABLE 3 Certified and recommended total elemental concentration values reported for plant and related reference materials listed in order of elements and in increasing order of concentration[a] (continued)

Material[b]	Code[b]	Concentration, mg/kg[c]
Ca: Calcium (continued)		
Rice flour	NIST-SRM-1568	140
Hard red spring wheat flour	NIST-RM-8437	143
Rye flour	IAEA-V-8	149
Wheat flour	NIST-SRM-1567	190
Wheat flour	NIST-SRM-1567A	191
Wheat flour	ARC/CL-WF	208
Rye flour	BCR-CRM-381	220
Rye bread flour	CSRM-12-2-05	234
Cotton cellulose	IAEA-V-9	240
Soft winter wheat flour	NIST-RM-8438	240
Durum wheat flour	NIST-RM-8436	278
Wheat bread flour	CSRM-12-2-04	292
Wheat gluten	NIST-RM-8418	369
Corn bran	NIST-RM-8433	420
Wheat flour	GBW-08503	441
Green algae (*C. kessleri*)	CSRM-12-2-02	539
Carrot powder	ARC/CL-CP	1700
Codonopsis P.	GBW-09501	1780
Corn stalk (*Zea mays*)	NIST-RM-8412	2160
Tea	GBW-08505	2840
Haricots verts (Beans)	BCR-CRM-383	2900
Tea leaves	NIES-CRM-7	3200
Pine needles	NIST-SRM-1575	4100
Spruce needles	BCR-CRM-101	4280
Tea	GBW-07605	4300
Chlorella	NIES-CRM-3	4900
Beech leaves	BCR-CRM-100	5300
Cabbage leaves	AMM-CL-1	6250
Hay powder	BCR-CRM-129	6400
Rice straw ash	DL-5702	7289
Cabbage	GBW-08504	7920
Single cell protein	BCR-CRM-273	11,970
Sargasso	NIES-CRM-9	13,400
Trace elements in spinach	NIST-SRM-1570	13,500
Pepperbush	NIES-CRM-1	13,800
Apple leaves	NIST-SRM-1515	15,260
Spinach leaves	NIST-SRM-1570A	15,270

TABLE 3 Certified and recommended total elemental concentration values reported for plant and related reference materials listed in order of elements and in increasing order of concentration[a] (continued)

Material[b]	Code[b]	Concentration, mg/kg[c]
Ca: Calcium (continued)		
Peach leaves	NIST-SRM-1547	15,600
Bush branches and leaves	GBW-07603	16,800
Lucerne P-alfalfa	CSRM-12-2-03	17,500
Poplar leaves	GBW-07604	18,100
Hay powder	IAEA-V-10	21,600
Bush branches and leaves	GBW-07602	22,200
Tomato leaves	NIST-SRM-1573	30,000
Citrus leaves	NIST-SRM-1572	31,500
Oriental tobacco leaves	ICHTJ-CTA-OTL-1	31,700
Kale	BOWEN'S KALE	41,060
Tomato leaves	NIST-SRM-1573A	50,500
Cd: Cadmium		
Microcrystalline cellulose	NIST-RM-8416	0.0002
Corn starch	NIST-RM-8432	0.0003
Rye bread flour	CSRM-12-2-05	0.0102
Corn bran	NIST-RM-8433	0.012
Apple leaves	NIST-SRM-1515	0.013
Peach leaves	GBW-08501	0.018
Rice flour	GBW-08502	0.020
Rice flour	NIST-SRM-1568A	0.022
Rice flour	NIES-CRM-10A	0.023
Wheat flour	NIST-SRM-1567A	0.026
Brown bread	BCR-CRM-191	0.0284
Cabbage	GBW-08504	0.029
Rice flour	NIST-SRM-1568	0.029
Single cell protein	BCR-CRM-274	0.030
Hay powder	IAEA-V-10	0.030
Tea leaves	NIST-SRM-7	0.030
Citrus leaves	NIST-SRM-1572	0.030
Wheat flour	GBW-08503	0.031
Tea	GBW-08505	0.032
Wheat flour	NIST-SRM-1567	0.032
Potato powder	ARC/CL-PP	0.035
Wheat flour	ARC/CL-WF	0.039
Wheat bread flour	CSRM-12-2-04	0.0415
Green algae (*C. kessleri*)	CSRM-12-2-02	0.0448

TABLE 3 Certified and recommended total elemental concentration values reported for plant and related reference materials listed in order of elements and in increasing order of concentration[a] (continued)

Material[b]	Code[b]	Concentration, mg/kg[c]
Cd: Cadmium (continued)		
Tea	GBW-07605	0.057
Wheat gluten	NIST-RM-8418	0.064
Carrot powder	ARC/CL-CP	0.0686
Whole meal flour	BCR-CRM-189	0.0713
Olive leaves (*O. europaea*)	BCR-CRM-062	0.10
Durum wheat flour	NIST-RM-8436	0.11
Lichen	IAEA-336	0.117
Rye grass	BCR-CRM-281	0.12
Yeast (Torulopsis E.)	CALNRI-TY-4	0.133
Lucerne P-alfalfa	CSRM-12-2-03	0.136
Bush branches and leaves	GBW-07602	0.14
Sargasso	NIES-CRM-9	0.15
Yeast (Torulopsis E.)	CALNRI-TY-2	0.212
Cabbage leaves	AMM-CL-1	0.23
Yeast (Torulopsis E.)	CALNRI-TY-1	0.264
Sea lettuce (*Ulva lactuca*)	BCR-CRM-279	0.274
Poplar leaves	GBW-07604	0.32
Rice flour	NIES-CRM-10B	0.32
Plankton	BCR-CRM-414	0.383
Cantharellus T. (fungus)	LIVSVER-FUNGUS	0.437
Lichen	BCR-CRM-482	0.56
Aquatic plant (*P. riparioides*)	BCR-CRM-061	1.07
Oriental tobacco Leaves	ICHTJ-CTA-OTL-1	1.12
Tomato leaves	NIST-SRM-1573A	1.52
Rice flour	NIST-CRM-10C	1.82
Aquatic plant (*L. major*)	BCR-CRM-060	2.2
Spinach leaves	NIST-SRM-1570A	2.89
Pepperbush	NIES-CRM-1	6.7
Ce: Cerium		
Poplar leaves	GBW-07604	0.49
Tea	GBW-08505	0.686
Tea	GBW-07605	1.0
Lichen	IAEA-336	1.27
Bush branches and leaves	GBW-07603	2.2
Bush branches and leaves	GBW-07602	2.4
Oriental tobacco leaves	ICHTJ-CTA-OTL-1	2.69

TABLE 3 Certified and recommended total elemental concentration values reported for plant and related reference materials listed in order of elements and in increasing order of concentration[a] (continued)

Material[b]	Code[b]	Concentration, mg/kg[c]
Cl: Chlorine		
Corn bran	NIST-RM-8433	31
Corn starch	NIST-RM-8432	45
Microcrystalline cellulose	NIST-RM-8416	80
Peach leaves	NIST-SRM-1547	360
Rye flour	BCR-CRM-381	460
Hard red spring wheat flour	NIST-RM-8437	500
Rye flour	IAEA-V-8	570
Apple leaves	NIST-SRM-1515	579
Cotton cellulose	IAEA-V-9	600
Soft winter wheat flour	NIST-RM-8438	640
Durum wheat flour	NIST-RM-8436	680
Spruce needles	BCR-CRM-101	688
Beech leaves	BCR-CRM-100	1490
Corn stalk (*Zea mays*)	NIST-RM-8412	2440
Kale	BOWEN'S KALE	3560
Wheat gluten	NIST-RM-8418	3620
Co: Cobalt		
Corn starch	NIST-RM-8432	0.0012
Microcrystalline cellulose	NIST-RM-8416	0.0017
Corn bran	NIST-RM-8433	0.006
Durum wheat flour	NIST-RM-8436	0.008
Wheat gluten	NIST-RM-8418	0.01
Rice flour	NIST-SRM-1568	0.02
Single cell protein	BCR-CRM-274	0.039
Kale	BOWEN'S KALE	0.0632
Cantharellus T. (Fungus)	LIVSVER-FUNGUS	0.073
Yeast (Torulopsis E.)	CALNRI-TY-1	0.115
Sargasso	NIES-CRM-9	0.12
Hay powder	IAEA-V-10	0.13
Yeast (Torulopsis E.)	CALNRI-TY-2	0.17
White clover	BCR-CRM-402	0.178
Tea	GBW-07605	0.18
Yeast (Torulopsis E.)	CALNRI-TY-3	0.199
Lichen	IAEA-336	0.287
Yeast (Torulopsis E.)	CALNRI-TY-4	0.298
Bush branches and leaves	GBW-07602	0.39
Spinach leaves	NIST-SRM-1570A	0.39

TABLE 3 Certified and recommended total elemental concentration values reported for plant and related reference materials listed in order of elements and in increasing order of concentration[a] (continued)

Material[b]	Code[b]	Concentration, mg/kg[c]
Co: Cobalt (continued)		
Bush branches and leaves	GBW-07603	0.41
Poplar leaves	GBW-07604	0.42
Tomato leaves	NIST-SRM-1573A	0.57
Chlorella	NIES-CRM-3	0.87
Oriental tobacco Leaves	ICHTJ-CTA-OTL-1	0.879
Green algae (*C. kessleri*)	CSRM-12-2-02	19.9
Pepperbush	NIES-CRM-1	23
Cr: Chromium		
Durum wheat flour	NIST-RM-8436	0.023
Hard red spring wheat flour	NIST-RM-8437	0.026
Wheat flour	ARC/CL-WF	0.028
Soft winter wheat flour	NIST-RM-8438	0.032
Wheat gluten	NIST-RM-8418	0.053
Corn bran	NIST-RM-8433	0.101
Cotton cellulose	IAEA-V-9	0.11
Cantharellus T. (Fungus)	LIVSVER-FUNGUS	0.223
Poplar leaves	GBW-07604	0.55
Tea	GBW-07605	0.8
Citrus leaves	NIST-SRM-1572	0.8
Peach leaves	GBW-08501	0.94
Tomato leaves	NIST-SRM-1573A	1.99
Brewer's yeast	NIST-SRM-1569	2.12
Bush branches and leaves	GBW-07602	2.3
Green algae (*C. kessleri*)	CSRM-12-2-02	2.37
Oriental tobacco leaves	ICHTJ-CTA-OTL-1	2.59
Bush branches and leaves	GBW-07603	2.6
Pine needles	NIST-SRM-1575	2.6
Lichen	BCR-CRM-482	4.12
Tomato leaves	NIST-SRM-1573	4.5
Trace elements in spinach	NIST-SRM-1570	4.6
Hay powder	IAEA-V-10	6.5
Plankton	BCR-CRM-414	23.8
Yeast (Torulopsis E.)	CALNRI-TY-2	32
Aquatic plant (*Trapa natans*)	BCR-CRM-596	36.3
Yeast (Torulopsis E.)	CALNRI-TY-3	87.8
Yeast (Torulopsis E.)	CALNRI-TY-4	221

TABLE 3 Certified and recommended total elemental concentration values reported for plant and related reference materials listed in order of elements and in increasing order of concentration[a] (continued)

Material[b]	Code[b]	Concentration, mg/kg[c]
Cs: Cesium		
Poplar leaves	GBW-07604	0.053
Kale	BOWEN'S KALE	0.0763
Lichen	IAEA-336	0.11
Oriental tobacco leaves	ICHTJ-CTA-OTL-1	0.177
Bush branches and leaves	GBW-07602	0.27
Bush branches and leaves	GBW-07603	0.27
Tea	GBW-07605	0.29
Cu: Copper		
Microcrystalline cellulose	NIST-RM-8416	0.015
Corn starch	NIST-RM-8432	0.06
Cotton cellulose	IAEA-V-9	0.59
Rye flour	IAEA-V-8	0.95
Soft winter wheat flour	NIST-RM-8438	1.2
Wheat flour	NIST-SRM-1567	2
Hard red spring wheat flour	NIST-RM-8437	2.01
Wheat flour	NIST-SRM-1567A	2.1
Rice flour	NIST-SRM-1568	2.2
Wheat flour	ARC/CL-WF	2.35
Rice flour	NIST-SRM-1568A	2.4
Rye bread flour	CSRM-12-2-05	2.43
Corn bran	NIST-RM-8433	2.47
Brown bread	BCR-CRM-191	2.6
Rice flour	GBW-08502	2.6
Wheat bread flour	CSRM-12-2-04	2.77
Cabbage	GBW-08504	3
Corn kernel (*Zea mays*)	NIST-RM-8413	3
Pine needles	NIST-SRM-1575	3
Cabbage leaves	AMM-CL-1	3.3
Rice flour	NIES-CRM-10B	3.3
Rice flour	NIES-CRM-10A	3.5
Chlorella	NIES-CRM-3	3.5
Lichen	IAEA-336	3.55
Mercury in rice	GBW-08508	3.6
Potato powder	ARC/CL-PP	3.9
Rice flour	NIES-CRM-10C	4.1
Durum wheat flour	NIST-RM-8436	4.3

TABLE 3 Certified and recommended total elemental concentration values reported for plant and related reference materials listed in order of elements and in increasing order of concentration[a] (continued)

Material[b]	Code[b]	Concentration, mg/kg[c]
Cu: Copper (continued)		
Wheat flour	GBW-08503	4.4
Tomato leaves	NIST-SRM-1573A	4.7
Kale	BOWEN'S KALE	4.89
Sargasso	NIES-CRM-9	4.9
Bush branches and leaves	GBW-07602	5.2
Apple leaves	NIST-SRM-1515	5.64
Wheat gluten	NIST-RM-8418	5.94
Whole meal flour	BCR-CRM-189	6.4
Bush branches and leaves	GBW-07603	6.6
Codonopsis P.	GBW-09501	6.8
Tea leaves	NIES-CRM-7	7
Lichen	BCR-CRM-482	7.03
Corn stalk (*Zea mays*)	NIST-RM-8412	8
Poplar leaves	GBW-07604	9.3
Hay powder	IAEA-V-10	9.4
Rye grass	BCR-CRM-281	9.65
Peach leaves	GBW-08501	10.4
Tomato leaves	NIST-SRM-1573	11
Lucerne P-alfalfa	CSRM-12-2-03	11.7
Pepperbush	NIES-CRM-1	12
Trace elements in spinach	NIST-SRM-1570	12
Spinach leaves	NIST-SRM-1570A	12.2
Single cell protein	BCR-CRM-274	13.1
Sea lettuce (*Ulva lactuca*)	BCR-CRM-279	13.14
Oriental tobacco Leaves	ICHTJ-CTA-OTL-1	14.1
Tea	GBW-08505	16.2
Citrus leaves	NIST-SRM-1572	16.5
Tea	GBW-07605	17.3
Green algae (*C. kessleri*)	CSRM-12-2-02	19.6
Plankton	BCR-CRM-414	29.5
Cantharellus T. (fungus)	LIVSVER-FUNGUS	34.4
Yeast (Torulopsis E.)	CALNRI-TY-4	42.1
Olive leaves (*O. europaea*)	BCR-CRM-062	46.6
Aquatic plant (*L. major*)	BCR-CRM-060	51.2
Yeast (Torulopsis E.)	CALNRI-TY-3	52.8
Yeast (Torulopsis E.)	CALNRI-TY-2	61
Yeast (Torulopsis E.)	CALNRI-TY-1	65.7

TABLE 3 Certified and recommended total elemental concentration values reported for plant and related reference materials listed in order of elements and in increasing order of concentration[a] (continued)

Material[b]	Code[b]	Concentration, mg/kg[c]
Cu: Copper (continued)		
Peach leaves	NIST-SRM-1547	307
Aquatic plant (*P. riparioides*)	BCR-CRM-061	720
Eu: Europium		
Poplar leaves	GBW-07604	0.009
Tea	GBW-07605	0.018
Bush branches and leaves	GBW-07602	0.037
Oriental tobacco leaves	ICHTJ-CTA-OTL-1	0.038
Bush branches and leaves	GBW-07603	0.039
F: Fluorine		
Fluoride composition in corn	GBW-08506	1.91
Kale	BOWEN'S KALE	5.87
Poplar leaves	GBW-07604	22
Bush branches and leaves	GBW-07603	23
Bush branches and leaves	GBW-07602	24
Fluoride composition in corn	GBW-08507	33.7
Fluoride in vegetation, low	NIST-SRM-2695 L	64
Fluoride in vegetation, high	NIST-SRM-2695 H	277
Tea	GBW-07605	320
Fe: Iron		
Rye flour	IAEA-V-8	4.1
Rice flour	GBW-08502	5.1
Rice flour	NIST-SRM-1568A	7.4
Rice flour	NIST-SRM-1568	8.7
Rice flour	NIES-CRM-10C	11.4
Carrot powder	ARC/CL-CP	12.6
Rice flour	NIES-CRM-10A	12.7
Rice flour	NIES-CRM-10B	13.4
Wheat flour	NIST-SRM-1567A	14.1
Corn bran	NIST-RM-8433	14.8
Wheat flour	NIST-SRM-1567	18.3
Rye bread flour	CSRM-12-2-05	20.8
Potato powder	ARC/CL-PP	22
Corn kernel (*Zea mays*)	NIST-RM-8413	23
Wheat bread flour	CSRM-12-2-04	23.8
Soft winter wheat flour	NIST-RM-8438	29

TABLE 3 **Certified and recommended total elemental concentration values reported for plant and related reference materials listed in order of elements and in increasing order of concentration[a] (continued)**

Material[b]	Code[b]	Concentration, mg/kg[c]
Fe: Iron (continued)		
Hard red spring wheat flour	NIST-RM-8437	31
Wheat flour	GBW-08503	39.8
Brown bread	BCR-CRM-191	40.7
Durum wheat flour	NIST-RM-8436	41.5
Mercury in rice	GBW-08508	43.2
Wheat flour	ARC/CL-WF	51
Cabbage	GBW-08504	52
Wheat gluten	NIST-RM-8418	54.3
Cabbage leaves	AMM-CL-1	58.4
Whole meal flour	BCR-CRM-189	68.3
Codonopsis P.	GBW-09501	69.5
Citrus leaves	NIST-SRM-1572	90
Cantharellus T. (fungus)	LIVSVER-FUNGUS	101
Kale	BOWEN'S KALE	119.3
Corn stalk (*Zea mays*)	NIST-RM-8412	139
Single cell protein	BCR-CRM-273	156
Yeast (Torulopsis E.)	CALNRI-TY-4	156
Yeast (Torulopsis E.)	CALNRI-TY-3	162
Hay powder	IAEA-V-10	185
Sargasso	NIES-CRM-9	187
Pine needles	NIST-SRM-1575	200
Pepperbush	NIES-CRM-1	205
Peach leaves	NIST-SRM-1547	218
Yeast (Torulopsis E.)	CALNRI-TY-2	240
Tea	GBW-07605	264
Poplar leaves	GBW-07604	274
Yeast (Torulopsis E.)	CALNRI-TY-1	300
Green algae (*C. kessleri*)	CSRM-12-2-02	339
Lucerne P-alfalfa	CSRM-12-2-03	355
Tomato leaves	NIST-SRM-1573A	368
Tea	GBW-08505	373
Lichen	IAEA-336	426
Peach leaves	GBW-08501	431
Trace elements in spinach	NIST-SRM-1570	550
Tomato leaves	NIST-SRM-1573	690
Bush branches and leaves	GBW-07602	1,020
Bush branches and leaves	GBW-07603	1,070

TABLE 3 Certified and recommended total elemental concentration values reported for plant and related reference materials listed in order of elements and in increasing order of concentration[a] (continued)

Material[b]	Code[b]	Concentration, mg/kg[c]
Fe: Iron (continued)		
Chlorella	NIES-CRM-3	1,850
Rice straw ash	DL-5702	5,357
Hf: Hafnium		
Bush branches and leaves	GBW-07602	0.14
Hg: Mercury		
Durum wheat flour	NIST-RM-8436	0.0004
Wheat flour	NIST-SRM-1567	0.001
Corn starch	NIST-RM-8432	0.0011
Wheat gluten	NIST-RM-8418	0.0019
Corn bran	NIST-RM-8433	0.003
Rice flour	NIST-SRM-1568A	0.0058
Rice flour	NIST-SRM-1568	0.006
Hay powder	IAEA-V-10	0.013
Rye grass	BCR-CRM-281	0.0205
Poplar leaves	GBW-07604	0.026
Yeast (Torulopsis E.)	CALNRI-TY-3	0.029
Trace elements in spinach	NIST-SRM-1570	0.03
Spinach leaves	NIST-SRM-1570A	0.03
Peach leaves	NIST-SRM-1547	0.031
Tomato leaves	NIST-SRM-1573A	0.034
Mercury in rice	GBW-08508	0.038
Yeast (Torulopsis E.)	CALNRI-TY-4	0.044
Apple leaves	NIST-SRM-1515	0.044
Peach leaves	GBW-08501	0.046
Yeast (Torulopsis E.)	CALNRI-TY-2	0.05
Cotton cellulose	IAEA-V-9	0.06
Yeast (Torulopsis E.)	CALNRI-TY-1	0.063
Citrus leaves	NIST-SRM-1572	0.08
Pine needles	NIST-SRM-1575	0.15
Kale	BOWEN'S KALE	0.171
Lichen	IAEA-33	0.2
Aquatic plant (*P. riparioides*)	BCR-CRM-061	0.23
Plankton	BCR-CRM-414	0.276
Olive leaves (*O. europaea*)	BCR-CRM-062	0.28
Aquatic plant (*L. major*)	BCR-CRM-060	0.34
Lichen	BCR-CRM-482	0.48

TABLE 3 Certified and recommended total elemental concentration values reported for plant and related reference materials listed in order of elements and in increasing order of concentration[a] (continued)

Material[b]	Code[b]	Concentration, mg/kg[c]
I: Iodine		
Durum wheat flour	NIST-RM-8436	0.006
Corn bran	NIST-RM-8433	0.026
Wheat gluten	NIST-RM-8418	0.06
Hay powder	BCR-CRM-129	0.167
Citrus leaves	NIST-SRM-1572	1.84
K: Potassium		
Corn starch	NIST-RM-8432	45
Wheat gluten	NIST-RM-8418	472
Corn bran	NIST-RM-8433	566
Rice flour	GBW-08502	656
Rice flour	NIST-SRM-1568	1,120
Hard red spring wheat flour	NIST-RM-8437	1,150
Rice flour	NIST-SRM-1568A	1,280
Wheat flour	NIST-SRM-1567A	1,330
Wheat flour	NIST-SRM-1567	1,360
Soft winter wheat flour	NIST-RM-8438	1,480
Lichen	IAEA-336	1,840
Rye flour	IAEA-V-8	1,925
Wheat flour	GBW-08503	1,980
Wheat flour	ARC/CL-WF	2,200
Single cell protein	BCR-CRM-273	2,220
Rice flour	NIES-CRM-10B	2,450
Wheat bread flour	CSRM-12-2-04	2,550
Rye bread flour	CSRM-12-2-05	2,670
Rice flour	NIES-CRM-10C	2,750
Rice flour	NIES-CRM-10A	2,800
Rye flour	BCR-CRM-381	2,900
Durum wheat flour	NIST-RM-8436	3,180
Corn kernel (*Zea mays*)	NIST-RM-8413	3,570
Pine needles	NIST-SRM-1575	3,700
Haricots verts (beans)	BCR-CRM-383	7,800
Bush branches and leaves	GBW-07602	8,500
Bush branches and leaves	GBW-07603	9,200
Beech leaves	BCR-CRM-100	9,940
Carrot powder	ARC/CL-CP	10,200
Chlorella	NIES-CRM-3	12,400
Rice straw ash	DL-5702	13,773

TABLE 3 Certified and recommended total elemental concentration values reported for plant and related reference materials listed in order of elements and in increasing order of concentration[a] (continued)

Material[b]	Code[b]	Concentration, mg/kg[c]
K: Potassium (continued)		
Poplar leaves	GBW-07604	13,800
Cabbage	GBW-08504	14,500
Pepperbush	NIES-CRM-1	15,100
Oriental tobacco leaves	ICHTJ-CTA-OTL-1	15,600
Apple leaves	NIST-SRM-1515	16,100
Tea	GBW-07605	16,600
Yeast (Torulopsis E.)	CALNRI-TY-4	16,700
Corn stalk (*Zea mays*)	NIST-RM-8412	17,350
Green algae (*C. kessleri*)	CSRM-12-2-02	18,100
Citrus leaves	NIST-SRM-1572	18,200
Yeast (Torulopsis E.)	CALNRI-TY-3	18,500
Tea leaves	NIES-CRM-7	18,600
Lucerne P-alfalfa	CSRM-12-2-03	18,700
Yeast (Torulopsis E.)	CALNRI-TY-2	19,100
Yeast (Torulopsis E.)	CALNRI-TY-1	19,600
Tea	GBW-08505	19,700
Peach leaves	GBW-08501	21,700
Peach leaves	NIST-SRM-1547	24,300
Kale	BOWEN'S KALE	24,370
Cabbage leaves	AMM-CL-1	26,200
Tomato leaves	NIST-SRM-1573A	27,000
Spinach leaves	NIST-SRM-1570A	29,030
Hay powder	BCR-CRM-129	33,800
Trace elements in spinach	NIST-SRM-1570	35,600
Tomato leaves	NIST-SRM-1573	44,600
Sargasso	NIES-CRM-9	61,000
La: Lanthanum		
Kale	BOWEN'S KALE	0.0864
Poplar leaves	GBW-07604	0.26
Tea	GBW-08505	0.458
Tea	GBW-07605	0.6
Bush branches and leaves	GBW-07602	1.23
Bush branches and leaves	GBW-07603	1.25
Oriental tobacco leaves	ICHTJ-CTA-OTL- 1	1.44
Li: Lithium		
Poplar leaves	GBW-07604	0.84

TABLE 3 Certified and recommended total elemental concentration values reported for plant and related reference materials listed in order of elements and in increasing order of concentration[a] (continued)

Material[b]	Code[b]	Concentration, mg/kg[c]
Li: Lithium (continued)		
Bush branches and leaves	GBW-07602	2.4
Bush branches and leaves	GBW-07603	2.6
Oriental tobacco leaves	ICHTJ-CTA-OTL-1	23
Mg: Magnesium		
Corn starch	NIST-RM-8432	31
Cotton cellulose	IAEA-V-9	53
Rice flour	GBW-08502	120
Rye flour	IAEA-V-8	121
Soft winter wheat flour	NIST-RM-8438	214
Carrot powder	ARC/CL-CP	350
Hard red spring wheat flour	NIST-RM-8437	365
Wheat flour	NIST-SRM-1567	400
Rye bread flour	CSRM-12-2-05	407
Rye flour	BCR-CRM-381	430
Wheat gluten	NIST-RM-8418	510
Wheat flour	GBW-08503	551
Wheat bread flour	CSRH-12-2-04	556
Rice flour	NIST-SRM-1568A	560
Wheat flour	ARC/CL-WF	562
Spruce needles	BCR-CRM-101	619
Potato powder	ARC/CL-PP	750
Corn bran	NIST-RM-8433	818
Beech leaves	BCR-CRM-100	878
Corn kernel (*Zea mays*)	NIST-RM-8413	990
Durum wheat flour	NIST-RM-8436	1,070
Rice flour	NIES-CRM-10C	1,250
Rice flour	NIES-CRM-10B	1,310
Rice flour	NIES-CRM-10A	1,340
Hay powder	IAEA-V-10	1,360
Hay powder	BCR-CRM-129	1,450
Tea leaves	NIES-CRM-7	1,530
Corn stalk (*Zea mays*)	NIST-RM-8412	1,600
Kale	BOWEN'S KALE	1,605
Tea	GRW-07605	1,700
Cabbage	GRW-08504	1,840
Cabbage leaves	AMM-CL-1	1,850
Tea	GRW-08505	2,240

TABLE 3 Certified and recommended total elemental concentration values reported for plant and related reference materials listed in order of elements and in increasing order of concentration[a] (continued)

Material[b]	Code[b]	Concentration, mg/kg[c]
Mg: Magnesium (continued)		
Yeast (Torulopsis E.)	CALNRI-TY-3	2,470
Single cell protein	BCR-CRM-273	2,700
Apple leaves	NIST-SRM-1515	2,710
Yeast (Torulopsis E.)	CALNRI-TY-2	2,860
Bush branches and leaves	GBW-07602	2,870
Rice straw ash	DL-5702	3,100
Yeast (Torulopsis E.)	CALNRI-TY-1	3,200
Chlorella	NIES-CRM-3	3,300
Lucerne P-alfalfa	CSRM-12-2-03	3,520
Pepperbush	NIES-CRM-1	4,080
Peach leaves	NIST-SRM-1547	4,320
Oriental tobacco leaves	ICHTJ-CTA-OTL-1	4,470
Peach leaves	GBW-08501	4,700
Bush branches and leaves	GBW-07603	4,800
Citrus leaves	NIST-SRM-1572	5,800
Poplar leaves	GBW-07604	6,500
Sargasso	NIES-CRM-9	6,500
Green algae (*C. kessleri*)	CSRM-12-2-02	10,900
Mn: Manganese		
Corn starch	NIST-RM-8432	0.1
Cotton cellulose	IAEA-V-9	0.15
Rye flour	IAEA-V-8	2.06
Corn bran	NIST-RM-8433	2.55
Yeast (Torulopsis E.)	CALNRI-TY-1	2.68
Yeast (Torulopsis E.)	CALNRI-TY-2	3.47
Yeast (Torulopsis E.)	CALNRI-TY-4	3.85
Corn kernel (*Zea mays*)	NIST-RM-8413	4
Hard red spring wheat flour	NIST-RM-8437	4.5
Yeast (Torulopsis E.)	CALNRI-TY-3	4.77
Carrot powder	ARC/CL-CP	4.9
Soft winter wheat flour	NIST-RM-8438	5.4
Potato powder	ARC/CL-PP	8.1
Wheat flour	NIST-SRM-1567	8.5
Wheat flour	NIST-SRM-1567A	9.4
Rice flour	GBW-08502	9.8
Wheat flour	ARC/CL-WF	12.87
Wheat gluten	NIST-RM-8418	14.3

TABLE 3 Certified and recommended total elemental concentration values reported for plant and related reference materials listed in order of elements and in increasing order of concentration[a] (continued)

Material[b]	Code[b]	Concentration, mg/kg[c]
Mn: Manganese (continued)		
Corn stalk (*Zea mays*)	NIST-RM-8412	15
Durum wheat flour	NIST-RM-8436	16
Rye bread flour	CSRM-12-2-05	18
Codonopsis P.	GBW-09501	18.8
Wheat flour	GBW-08503	19.6
Rice flour	NIST-SRM-1568A	20
Brown bread	BCR-CRM-191	20.3
Rice flour	NIST-SRM-1568	21
Sargasso	NIES-CRM-9	21.2
Cabbage	GBW-08504	22
Wheat bread flour	CSRM-12-2-04	22.6
Citrus leaves	NIST-SRM-1572	23
Mercury in rice	GBW-08508	28.4
Rice flour	NIES-CRM-10B	31.5
Green algae (*C. kessleri*)	CSRM-12-2-02	32.8
Lucerne P-Alfalfa	CSRM-12-2-03	34.2
Rice flour	NIES-CRM-10A	34.7
Rice flour	NIES-CRM-10C	40.1
Poplar leaves	GBW-07604	45
Cantharellus T. (fungus)	LIVSVER-FUNGUS	49.9
Single cell protein	BCR-CRM-274	51.9
Apple leaves	NIST-SRM-1515	54
Olive leaves (*O. europaea*)	BCR-CRM-062	57
Cabbage leaves	AMM-CL-1	57.6
Bush branches and leaves	GBW-07602	58
Bush branches and leaves	GBW-07603	61
Whole meal flour	BCR-CRM-189	63.3
Lichen	IAEA-336	64
Chlorella	NIES-CRM-3	69
Peach leaves	GBW-08501	75.4
Spinach leaves	NIST-SRM-1570A	75.9
Rye grass	BCR-CRM-281	81.6
Peach leaves	NIST-SRM-1547	98
Trace elements in spinach	NIST-SRM-1570	165
Tomato leaves	NIST-SRM-1573	238
Tomato leaves	NIST-SRM-1573A	246
Plankton	BCR-CRM-414	299
Oriental tobacco leaves	ICHTJ-CTA-OTL-1	412

TABLE 3 Certified and recommended total elemental concentration values reported for plant and related reference materials listed in order of elements and in increasing order of concentration[a] (continued)

Material[b]	Code[b]	Concentration, mg/kg[c]
Mn: Manganese (continued)		
Rice straw ash	DL-5702	439
Pine needles	NIST-SRM-1575	675
Tea leaves	NIES-SRM-7	700
Tea	GBW-08505	766
Spruce needles	BCR-CRM-101	915
Tea	GBW-07605	1,240
Aquatic plant (*L. major*)	BCR-CRM-060	1,759
Pepperbush	NIES-CRM-1	2,030
Aquatic plant (*P. riparioides*)	BCR-CRM-061	3,771
Mo: Molybdenum		
Microcrystalline cellulose	NIST-RM-8416	0.01
Corn starch	NIST-RM-8432	0.02
Cotton cellulose	IAEA-V-9	0.034
Tea	GBW-07605	0.038
Carrot powder	ARC/CL-CP	0.0495
Peach leaves	NIST-SRM-1547	0.06
Apple leaves	NIST-SRM-1515	0.094
Citrus leaves	NIST-SRM-1572	0.17
Poplar leaves	GBW-07604	0.18
Potato powder	ARC/CL-PP	0.21
Wheat flour	ARC/CL-WF	0.25
Corn bran	NIST-RM-8433	0.252
Bush branches and leaves	GBW-07602	0.26
Bush branches and leaves	GBW-07603	0.28
Soft winter wheat flour	NIST-RM-8438	0.29
Rice flour	NIES-CRM-10A	0.35
Rice flour	NIES-CRM-10B	0.42
Wheat flour	NIST-SRM-1567A	0.48
Hard red spring wheat flour	NIST-RM-8437	0.55
Durum wheat flour	NIST-RM-8436	0.7
Wheat gluten	NIST-RM-8418	0.76
Rye grass	BCR-CRM-281	0.84
Hay powder	IAEA-V-10	0.9
Rice flour	NIST-SRM-1568A	1.46
Rice flour	NIES-CRM-10C	1.6
Kale	BOWEN'S KALE	2.27
White clover	BCR-CRM-402	6.93

TABLE 3 Certified and recommended total elemental concentration values reported for plant and related reference materials listed in order of elements and in increasing order of concentration[a] (continued)

Material[b]	Code[b]	Concentration, mg/kg[c]
N: Nitrogen		
Microcrystalline cellulose	NIST-RM-8416	200
Corn starch	NIST-RM-8432	670
Corn bran	NIST-RM-8433	8,840
Haricots verts (beans)	BCR-CRM-383	10,500
Bush branches and leaves	GBW-07602	12,000
Rye flour	BCR-CRM-381	12,500
Bush branches and leaves	GBW-07603	15,000
Soft winter wheat flour	NIST-RM-8438	17,560
Spruce needles	BCR-CRM-101	18,890
Wheat flour	BCR-CRM-382	21,200
Apple leaves	NIST-SRM-1515	22,500
Poplar leaves	GBW-07604	25,600
Beech leaves	BCR-CRM-100	26,290
Hard red spring wheat flour	NIST-RM-8437	26,900
Durum wheat flour	NIST-RM-8436	27,070
Cabbage	GBW-08504	28,000
Peach leaves	NIST-SRM-1547	29,400
Tomato leaves	NIST-SRM-1573A	30,300
Tea	GBW-07605	33,200
Hay powder	BCR-CRM-129	37,200
Kale	BOWEN'S KALE	42,790
Tea	GBW-08505	48,800
Spinach leaves	NIST-SRM-1570A	59,000
Wheat gluten	NIST-RM-8418	146,800
Na: Sodium		
Rice flour	NIST-SRM-1568	6
Wheat flour	NIST-SRM-1567A	6.1
Rice flour	NIST-SRM-1568A	6.6
Hard red spring wheat flour	NIST-RM-8437	7
Soft winter wheat flour	NIST-RM-8438	7
Wheat flour	NIST-SRM-1567	8
Rice flour	GBW-08502	8.4
Rice flour	NIES-CRM-10A	10.2
Rice flour	NIES-CRM-10C	14
Tea leaves	NIES-CRM-7	15.5
Durum wheat flour	NIST-RM-8436	16
Rice flour	NIES-CRM-10B	17.8

TABLE 3 Certified and recommended total elemental concentration values reported for plant and related reference materials listed in order of elements and in increasing order of concentration[a] (continued)

Material[b]	Code[b]	Concentration, mg/kg[c]
Na: Sodium (continued)		
Rye flour	BCR-CRM-381	19
Peach leaves	NIST-SRM-1547	24
Apple leaves	NIST-SRM-1515	24.4
Corn stalk (*Zea mays*)	NIST-RM-8412	28
Tea	GBW-07605	44
Cotton cellulose	IAEA-V-9	56
Haricots verts (beans)	BCR-CRM-383	75
Pepperbush	NIES-CRM-1	106
Corn starch	NIST-RM-8432	119
Tomato leaves	NIST-SRM-1573A	136
Tea	GBW-08505	142
Citrus leaves	NIST-SRM-1572	160
Green algae (*C. kessleri*)	CSRM-12-2-02	163
Poplar leaves	GBW-07604	200
Corn bran	NIST-RM-8433	430
Lucerne P-alfalfa	CSRM-12-2-03	474
Carrot powder	ARC/CL-CP	485
Yeast (Torulopsis E.)	CALNRI-TY-1	745
Yeast (Torulopsis E.)	CALNRI-TY-2	1,140
Wheat gluten	NIST-RM-8418	1,420
Yeast (Torulopsis E.)	CALNRI-TY-3	1,720
Rice straw ash	DL-5702	2,218
Kale	BOWEN'S KALE	2,366
Yeast (Torulopsis E.)	CALNRI-TY-4	2,580
Cabbage leaves	AMM-CL-1	3,030
Cabbage	GRW-08504	7,570
Bush branches and leaves	GRW-07602	11,000
Sargasso	NIES-CRM-9	17,000
Spinach leaves	NIST-SRM-1570A	18,180
Bush branches and leaves	GBW-07603	19,600
Nd: Neodymium		
Bush branches and leaves	GBW-07603	1
Ni: Nickel		
Corn starch	NIST-RM-8432	0.02
Microcrystalline cellulose	NIST-RM-8416	0.05
Cotton cellulose	IAEA-V-9	0.09
Wheat gluten	NIST-RM-8418	0.13

TABLE 3 Certified and recommended total elemental concentration values reported for plant and related reference materials listed in order of elements and in increasing order of concentration[a] (continued)

Material[b]	Code[b]	Concentration, mg/kg[c]
Ni: Nickel (continued)		
Wheat flour	ARC/CL-WF	0.153
Corn bran	NIST-RM-8433	0.158
Durum wheat flour	NIST-RM-8436	0.17
Rice flour	NIES-CRM-10A	0.19
Potato powder	ARC/CL-PP	0.193
Rice flour	NIES-CRM-10C	0.3
Cantharellus T. (fungus)	LIVSVER-FUNGUS	0.381
Rice flour	NIES-CRM-10B	0.39
Citrus leaves	NIST-SRM-1572	0.6
Peach leaves	NIST-SRM-1547	0.69
Yeast (Torulopsis E.)	CALNRI-TY-3	0.75
Yeast (Torulopsis E.)	CALNRI-TY-2	0.9
Apple leaves	NIST-SRM-1515	0.91
Yeast (Torulopsis E.)	CALNRI-TY-1	0.93
Tomato leaves	NIST-SRM-1573A	1.59
Bush branches and leaves	GBW-07602	1.7
Bush branches and leaves	GBW-07603	1.7
Poplar leaves	GBW-07604	1.9
Spinach leaves	NIST-SRM-1570A	2.14
Lichen	BCR-CRM-482	2.47
Lucerne P-alfalfa	CSRM-12-2-03	2.54
Cabbage leaves	AMM-CL-1	3
Rye grass	BCR-CRM-281	3
Hay powder	IAEA-V-10	4
Tea	GBW-07605	4.6
Oriental tobacco leaves	ICHTJ-CTA-OTL-1	6.32
Tea leaves	NIES-CRM-7	6.5
Tea	GBW-08505	7.61
Pepperbush	NIES-CRM-1	8.7
Plankton	BCR-CRM-414	18.8
P: Phosphorus		
Corn bran	NIST-RM-8433	171
Corn starch	NIST-RM-8432	178
Rye flour	IAEA-V-8	592
Bush branches and leaves	GBW-07602	830
Bush branches and leaves	GBW-07603	1,000
Soft winter wheat flour	NIST-RM-8438	1,080

TABLE 3 Certified and recommended total elemental concentration values reported for plant and related reference materials listed in order of elements and in increasing order of concentration[a] (continued)

Material[b]	Code[b]	Concentration, mg/kg[c]
P: Phosphorus (continued)		
Pine needles	NIST-SRM-1575	1,200
Citrus leaves	NIST-SRM-1572	1,300
Wheat flour	NIST-SRM-1567A	1,340
Hard red spring wheat flour	NIST-RM-8437	1,370
Rice flour	NIST-SRM-1568A	1,530
Beech leaves	BCR-CRM-100	1,550
Apple leaves	NIST-SRM-1515	1,590
Poplar leaves	GBW-07604	1,680
Spruce needles	BCR-CRM-101	1,690
Rye bread flour	CSRM-12-2-05	1,710
Wheat flour	ARC/CL-WF	2,090
Tomato leaves	NIST-SRM-1573A	2,160
Wheat gluten	NIST-RM-8418	2,190
Wheat bread flour	CSRM-12-2-04	2,280
Hay powder	IAEA-V-10	2,300
Hay powder	BCR-CRM-129	2,360
Tea	GBW-07605	2,840
Rice straw ash	DL-5702	2,875
Oriental tobacco leaves	ICHTJ-CTA-OTL-1	2,892
Durum wheat flour	NIST-RM-8436	2,900
Lucerne P-Alfalfa	CSRM-12-2-03	3,030
Rice flour	NIES-CRM-10B	3,150
Rice flour	NIES-CRM-10C	3,350
Cabbage	GBW-08504	3,400
Rice flour	NIES-CRM-10A	3,400
Tomato leaves	NIST-SRM-1573	3,400
Tea	GBW-08505	4,260
Kale	BOWEN'S KALE	4,480
Spinach leaves	NIST-SRM-1570A	5,180
Trace elements in spinach	NIST-SRM-1570	5,500
Yeast (Torulopsis E.)	CALNRI-TY-4	16,400
Green algae (*C. kessleri*)	CSRM-12-2-02	17,300
Yeast (Torulopsis E.)	CALNRI-TY-3	18,000
Yeast (Torulopsis E.)	CALNRI-TY-2	19,500
Yeast (Torulopsis E.)	CALNRI-TY-1	21,400
Single cell protein	BCR-CRM-273	26,800

TABLE 3 Certified and recommended total elemental concentration values reported for plant and related reference materials listed in order of elements and in increasing order of concentration[a] (continued)

Material[b]	Code[b]	Concentration, mg/kg[c]
Pb: Lead		
Microcrystalline cellulose	NIST-RM-8416	0.006
Corn starch	NIST-RM-8432	0.007
Wheat flour	ARC/CL-WF	0.018
Durum wheat flour	NIST-RM-8436	0.023
Potato powder	ARC/CL-PP	0.026
Wheat bread flour	CSRM-12-2-04	0.0414
Single cell protein	BCR-CRM-274	0.044
Rice flour	NIST-SRM-1568	0.045
Rye bread flour	CSRM-12-2-05	0.091
Wheat gluten	NIST-RM-8418	0.1
Corn bran	NIST-RM-8433	0.14
Brown bread	BCR-CRM-191	0.187
Cotton cellulose	IAEA-V-9	0.25
Cabbage leaves	AMM-CL-1	0.26
Cabbage	GBW-08504	0.28
Wheat flour	GBW-08503	0.35
Whole meal flour	BCR-CRM-189	0.379
Apple leaves	NIST-SRM-1515	0.47
Tea leaves	NIES-CRM-7	0.8
Peach leaves	NIST-SRM-1547	0.87
Peach leaves	GBW-08501	0.99
Tea	GBW-08505	1.06
Trace elements in spinach	NIST-SRM-1570	1.2
Green algae (*C. kessleri*)	CSRM-12-2-02	1.23
Sargasso	NIES-CRM-9	1.35
Cantharellus T. (fungus)	LIVSVER-FUNGUS	1.43
Poplar leaves	GBW-07604	1.5
Hay powder	IAEA-V-10	1.6
Lucerne P-Alfalfa	CSRM-12-2-03	1.84
Rye grass	BCR-CRM-281	2.38
Plankton	BCR-CRM-414	3.97
Tea	GBW-07605	4.4
Oriental tobacco leaves	ICHTJ-CTA-OTL-1	4.91
Pepperbush	NIES-CRM-1	5.5
Tomato leaves	NIST-SRM-1573	6.3
Bush branches and leaves	GBW-07602	7.1
Pine needles	NIST-SRM-1575	10.8

TABLE 3 Certified and recommended total elemental concentration values reported for plant and related reference materials listed in order of elements and in increasing order of concentration[a] (continued)

Material[b]	Code[b]	Concentration, mg/kg[c]
Pb: Lead (continued)		
Citrus leaves	NIST-SRM-1572	13.3
Sea lettuce (*Ulva lactuca*)	BCR-CRM-279	13.48
Olive leaves (*O. europaea*)	BCR-CRM-062	25
Lichen	BCR-CRM-482	40.9
Bush branches and leaves	GBW-07603	47
Aquatic plant (*L. major*)	BCR-CRM-060	63.8
Aquatic plant (*P. riparioides*)	BCR-CRM-061	64.4
Rb: Rubidium		
Rye flour	IAEA-V-8	0.48
Corn bran	NIST-RM-8433	0.5
Wheat flour	NIST-SRM-1567A	0.68
Lichen	IAEA-336	1.72
Durum wheat flour	NIST-RM-8436	2
Rice flour	NIES-CRM-10B	3.3
Bush branches and leaves	GBW-07602	4.2
Bush branches and leaves	GBW-07603	4.5
Rice flour	NIES-CRM-10A	4.5
Citrus leaves	NIST-SRM-1572	4.84
Yeast (Torulopsis E.)	CALNRI-TY-1	4.91
Rice flour	NIES-CRM-10C	5.7
Yeast (Torulopsis E.)	CALNRI-TY-2	5.82
Rice flour	NIST-SRM-1568A	6.14
Yeast (Torulopsis E.)	CALNRI-TY-3	6.93
Yeast (Torulopsis E.)	CALNRI-TY-4	7.43
Poplar leaves	GBW-07604	7.6
Hay powder	IAEA-V-10	7.6
Oriental tobacco leaves	ICHTJ-CTA-OTL-1	9.79
Apple leaves	NIST-SRM-1515	10.2
Pine needles	NIST-SRM-1575	11.7
Trace elements in spinach	NIST-SRM-1570	12.1
Tomato leaves	NIST-SRM-1573A	14.89
Tomato leaves	NIST-SRM-1573	16.5
Peach leaves	NIST-SRM-1547	19.7
Sargasso	NIES-CRM-9	24
Tea	GBW-08505	36.9
Cabbage leaves	AMM-CL-1	41.4
Kale	BOWEN'S KALE	53.4

TABLE 3   Certified and recommended total elemental concentration values reported for plant and related reference materials listed in order of elements and in increasing order of concentration[a] (continued)

Material[b]	Code[b]	Concentration, mg/kg[c]
Ru: Rubidium (continued)		
Tea	GBW-07605	74
Pepperbush	NIES-CRM-1	75
S: Sulfur		
Corn bran	NIST-RM-8433	860
Rice flour	NIST-SRM-1568A	1,200
Soft winter wheat flour	NIST-RM-8438	1,260
Wheat flour	NIST-SRM-1567A	1,650
Spruce needles	BCR-CRM-101	1,700
Hard red spring wheat flour	NIST-RM-8437	1,830
Durum wheat flour	NIST-RM-8436	1,930
Tea	GBW-07605	2,450
Beech leaves	BCR-CRM-100	2,690
Tea	GBW-08505	3,150
Hay powder	BCR-CRM-129	3,160
Bush branches and leaves	GBW-07602	3,200
Whole rapeseed (low level)	BCR-CRM-366	3,410
Poplar leaves	GBW-07604	3,500
Citrus leaves	NIST-SRM-1572	4,070
Whole rapeseed (medium level)	BCR-CRM-190	4,930
Bush branches and leaves	GBW-07603	7,300
Oriental tobacco leaves	ICHTJ-CTA-OTL-1	7,320
Wheat gluten	NIST-RM-8418	8,450
Whole rapeseed (high level)	BCR-CRM-367	10,510
Kale	BOWEN'S KALE	15,660
Sb: Antimony		
Tea	GBW-08505	0.037
Poplar leaves	GBW-07604	0.045
Rye grass	BCR-CRM-281	0.047
Tea	GBW-07605	0.056
Tomato leaves	NIST-SRM-1573A	0.063
Lichen	IAEA-336	0.073
Bush branches and leaves	GBW-07602	0.078
Yeast (Torulopsis E.)	CALNRI-TY-4	0.089
Bush branches and leaves	GBW-07603	0.095
Yeast (Torulopsis E.)	CALNRI-TY-3	0.102
Yeast (Torulopsis E.)	CALNRI-TY-2	0.124
Yeast (Torulopsis E.)	CALNRI-TY-1	0.137

TABLE 3 Certified and recommended total elemental concentration values reported for plant and related reference materials listed in order of elements and in increasing order of concentration[a] (continued)

Material[b]	Code[b]	Concentration, mg/kg[c]
Sc: Scandium		
Hay powder	IAEA-V-10	0.014
Poplar leaves	GBW-07604	0.069
Tea	GBW-07605	0.085
Lichen	IAEA-336	0.17
Bush branches and leaves	GBW-07602	0.31
Bush branches and leaves	GBW-07603	0.32
Se: Selenium		
Corn starch	NIST-RM-8432	0.0009
Microcrystalline cellulose	NIST-RM-8416	0.002
Corn kernel (*Zea mays*)	NIST-RM-8413	0.004
Corn stalk (*Zea mays*)	NIST-RM-8412	0.016
Rye grass	BCR-CRM-281	0.028
Tea	GBW-08505	0.041
Rice flour	GBW-08502	0.045
Corn bran	NIST-RM-8433	0.045
Apple leaves	NIST-SRM-1515	0.05
Tomato leaves	NIST-SRM-1573A	0.054
Wheat flour	ARC/CL-WF	0.057
Soft winter wheat flour	NIST-RM-8438	0.076
Cabbage	GBW-08504	0.083
Spinach leaves	NIST-SRM-1570A	0.117
Bush branches and leaves	GBW-07603	0.12
Peach leaves	NIST-SRM-1547	0.12
Whole meal flour	BCR-CRM-189	0.132
Poplar leaves	GBW-07604	0.14
Oriental tobacco leaves	ICHTJ-CTA-0TL-1	0.153
Bush branches and leaves	GBW-07602	0.184
Cabbage leaves	AMM-CL-1	0.2
Lichen	IAEA-336	0.216
Rice flour	NIST-SRM-1568A	0.38
Rice flour	NIST-SRM-1568	0.4
Hard red spring wheat flour	NIST-RM-8437	0.56
Sea lettuce (*Ulva lactuca*)	BCR-CRM-279	0.593
Single cell protein	BCR-CRM-274	1.03
Wheat flour	NIST-SRM-1567	1.1
Wheat flour	NIST-SRM-1567A	1.1
Durum wheat flour	NIST-RM-8436	1.23

TABLE 3 Certified and recommended total elemental concentration values reported for plant and related reference materials listed in order of elements and in increasing order of concentration[a] (continued)

Material[b]	Code[b]	Concentration, mg/kg[c]
Se: Selenium (continued)		
Plankton	BCR-CRM-414	1.75
Wheat gluten	NIST-RM-8418	2.58
White clover	BCR-CRM-402	6.7
Si: Silicon		
Bush branches and leaves	GBW-07602	5,800
Bush branches and leaves	GBW-07603	6,000
Poplar leaves	GBW-07604	7,100
Rice straw ash	DL-5702	440,338
Sm: Samarium		
Poplar leaves	GBW-07604	0.038
Tea	GBW-07605	0.085
Lichen	IAEA-336	0.106
Bush branches and leaves	GBW-07602	0.19
Bush branches and leaves	GBW-07603	0.19
Oriental tobacco leaves	ICHTJ-CTA-OTL-1	0.229
Sr: Strontium		
Corn starch	NIST-SRM-8432	0.18
Cotton cellulose	IAEA-V-9	0.65
Rye bread flour	CSRM-12-2-05	1.19
Durum wheat flour	NIST-RM-8436	1.19
Wheat gluten	NIST-RM-8418	1.71
Corn bran	NIST-RM-8433	4.62
Pine needles	NIST-SRM-1575	4.8
Tea	GBW-08505	10.8
Corn stalk (*Zea mays*)	NIST-RM-8412	12
Tea	GBW-07605	15.2
Codonopsis P.	GBW-09501	18
Cabbage leaves	AMM-CL-1	20.6
Apple leaves	NIST-SRM-1515	25
Pepperbush	NIES-CRM-1	36
Hay powder	IAEA-V-10	40
Chlorella	NIES-CRM-3	40
Tomato leaves	NIST-SRM-1573	44.9
Cabbage	GBW-08504	45.2
Peach leaves	NIST-SRM-1547	53
Spinach leaves	NIST-SRM-1570A	55.6

TABLE 3 Certified and recommended total elemental concentration values reported for plant and related reference materials listed in order of elements and in increasing order of concentration[a] (continued)

Material[b]	Code[b]	Concentration, mg/kg[c]
Sr: Strontium (continued)		
Peach leaves	GBW-08501	61.6
Lucerne P-alfalfa	CSRM-12-2-03	78.7
Trace elements in spinach	NIST-SRM-1570	87
Citrus leaves	NIST-SRM-1572	100
Poplar leaves	GBW-07604	154
Oriental tobacco leaves	ICHTJ-CTA-OTL-1	201
Bush branches and leaves	GBW-07603	246
Bush branches and leaves	GBW-07602	345
Sargasso	NIES-CRM-9	1,000
Tb: Terbium		
Bush branches and leaves	GBW-07603	0.025
Oriental tobacco leaves	ICHTJ-CTA-OTL-1	0.032
Th: Thorium		
Pine needles	NIST-SRM-1575	0.037
Grass	IAEA-373	0.047
Spinach leaves	NIST-SRM-1570A	0.048
Tea	GBW-07605	0.061
Poplar leaves	GBW-07604	0.07
Tea	GBW-08505	0.105
Trace elements in spinach	NIST-SRM-1570	0.12
Tomato leaves	NIST-SRM-1573	0.17
Oriental tobacco leaves	ICHTJ-CTA-OTL-1	0.348
Bush branches and leaves	GBW-07603	0.36
Bush branches and leaves	GBW-07602	0.37
Ti: Titanium		
Poplar leaves	GBW-07604	20.4
Tea	GBW-07605	24
Bush branches and leaves	GBW-07602	95
Bush branches and leaves	GBW-07603	95
Rice straw ash	DL-5702	1,163
U: Uranium		
Pine needles	NIST-SRM-1575	0.02
Trace elements in spinach	NIST-SRM-1570	0.046
Tomato leaves	NIST-SRM-1573	0.061
Spruce twigs and needles	CANMET-CLV-2	3.6
Spruce twigs and needles	CANMET-CLV-1	86.8

TABLE 3 Certified and recommended total elemental concentration values reported for plant and related reference materials listed in order of elements and in increasing order of concentration[a] (continued)

Material[b]	Code[b]	Concentration, mg/kg[c]
V: Vanadium		
Corn bran	NIST-RM-8433	0.005
Hard red spring wheat flour	NIST-RM-8437	0.02
Durum wheat flour	NIST-RM-8436	0.021
Apple leaves	NIST-SRM-1515	0.26
Peach leaves	NIST-SRM-1547	0.37
Kale	BOWEN'S KALE	0.386
Spinach leaves	NIST-SRM-1570A	0.57
Tomato leaves	NIST-SRM-1573A	0.835
Sargasso	NIES-CRM-9	1
Bush branches and leaves	GBW-07602	2.4
Bush branches and leaves	GBW-07603	2.4
Oriental tobacco leaves	ICHTJ-CTA-OTL-1	3.08
Plankton	BCR-CRM-414	8.1
Y: Yttrium		
Poplar leaves	GBW-07604	0.145
Tea	GBW-07605	0.36
Bush branches and leaves	GBW-07603	0.68
Yb: Ytterbium		
Poplar leaves	GBW-07604	0.018
Tea	GBW-07605	0.044
Bush branches and leaves	GBW-07602	0.063
Bush branches and leaves	GBW-07603	0.063
Zn: Zinc		
Corn starch	NIST-RM-8432	0.22
Rye flour	IAEA-V-8	2.53
Soft winter wheat flour	NIST-RM-8438	5.8
Carrot powder	ARC/CL-CP	6.9
Potato powder	ARC/CL-PP	9
Hard red spring wheat flour	NIST-RM-8437	10.6
Wheat flour	NIST-SRM-1567	10.6
Wheat flour	NIST-SRM-1567A	11.6
Apple leaves	NIST-SRM-1515	12.5
Rice flour	GBW-08502	14.1
Wheat flour	ARC/CL-WF	14.6
Codonopsis p.	GBW-09501	15.2
Sargasso	NIES-CRM-9	15.6

TABLE 3 Certified and recommended total elemental concentration values reported for plant and related reference materials listed in order of elements and in increasing order of concentration[a] (continued)

Material[b]	Code[b]	Concentration, mg/kg[c]
Zn: Zinc (continued)		
Corn kernel (*Zea mays*)	NIST-RM-8413	15.7
Olive leaves (*O. europaea*)	BCR-CRM-062	16
Rye bread flour	CSRM-12-2-05	16
Wheat bread flour	CSRM-12-2-04	17.9
Peach leaves	NIST-SRM-1547	17.9
Mercury in rice	GBW-08508	18
Corn bran	NIST-RM-8433	18.6
Rice flour	NIST-SRM-1568	19.4
Rice flour	NIST-SRM-1568A	19.4
Brown bread	BCR-CRM-191	19.5
Chlorella	NIES-CRM-3	20.5
Bush branches and leaves	GBW-07602	20.6
Durum wheat flour	NIST-RM-8436	22.2
Rice flour	NIES-RM-10B	22.3
Wheat flour	GBW-08503	22.7
Peach leaves	GBW-08501	22.8
Rice flour	NIES-CRM-10C	23.1
Hay powder	IAEA-V-10	24
Rice flour	NIES-CRM-10A	25.2
Tea	GBW-07605	26.3
Cabbage	GBW-08504	26.7
Citrus leaves	NIST-SRM-1572	29
Tomato leaves	NIST-SRM-1573A	30.9
Rye grass	BCR-CRM-281	31.5
Lichen	IAEA-336	31.6
Corn stalk (*Zea mays*)	NIST-RM-8412	32
Hay powder	BCR-CRM-129	32.1
Kale	BOWEN'S KALE	32.29
Tea leaves	NIES-CRM-7	33
Lucerne P-alfalfa	CSRM-12-2-03	33.2
Spruce needles	BCR-CRM-101	35.3
Poplar leaves	GBW-07604	37
Cabbage leaves	AMM-CL-1	38.5
Tea	GBW-08505	38.7
Green algae (*C. kessleri*)	CSRM-12-2-02	40.2
Single cell protein	BCR-CRM-274	42.7
Oriental tobacco leaves	ICHTJ-CTA-OTL-1	49.9

TABLE 3   Certified and recommended total elemental concentration values reported for plant and related reference materials listed in order of elements and in increasing order of concentration[a] (continued)

Material[b]	Code[b]	Concentration, mg/kg[c]
Zn: Zinc (continued)		
Trace elements in spinach	NIST-SRM-1570	50
Sea lettuce (*Ulva lactuca*)	BCR-CRM-279	51.3
Wheat gluten	NIST-RM-8418	53.8
Bush branches and leaves	GBW-07603	55
Cantharellus T. (fungus)	LIVSVER-FUNGUS	55
Whole meal flour	BCR-CRM-189	56.5
Tomato leaves	NIST-SRM-1573	62
Spinach leaves	NIST-SRM-1570A	82
Lichen	BCR-CRM-482	100.6
Plankton	BCR-CRM-414	112
Yeast (Torulopsis E.)	CALNRI-TY-3	226
Yeast (Torulopsis E.)	CALNRI-TY-4	266
Aquatic plant (*L. major*)	BCR-CRM-060	313
Yeast (Torulopsis E.)	CALNRI-TY-2	328
Pepperbush	NIES-CRM-1	340
Yeast (Torulopsis E.)	CALNRI-TY-1	383
Aquatic plant (*P. riparioides*)	BCR-CRM-061	566

[a] Based on information in the references listed at the end of the appendix including compilations of: International Atomic Energy Agency (1995), Cantillo (1995), Cortes Toro et al. (1990), Muramatsu and Parr (1985), and Ihnat (1988), as well as original certificates of analysis.

[b] Source and material codes are defined in Tables 1 and 2.

[c] Only those elements with total concentration values are listed. Information in this and other tables in this appendix should be used only as guides for Reference Material selection and use. It is emphasized that the latest appropriate certificates or reports of analysis or other relevant publications issued with the reference material when acquired must be consulted and used for actual data, significant figures, and other information. The most important source of information for the Reference Material is the certificate or report of analysis issued with the material. This document is an integral part of the Reference Material technology as it provides the analytical (certified) information, estimates of uncertainties, instructions for the correct use of the material, and other relevant information.

TABLE 4 Example of elemental concentration values from the report of investigation for reference material corn bran (NIST RM 8433)[a]

	Best estimate concentrations of constituent elements (dry weight basis)	
Element (major)	*Content and uncertainty (%)[b]*	*Methods[c]*
Nitrogen	0.884 ± 0.036	I01 I02 J01 J02

Element (minor and trace)	*Content and uncertainty (mg/kg)[b]*	*Methods[c]*
Sulfur	860 ± 150	B02 B03 J02 J03
Magnesium	818 ± 59	A01 A03 B02 B03 D01
Potassium	566 ± 75	A01 B01 B02 B03 D01 E01
Sodium	430 ± 31	A01 B01 B02 B03 D01
Calcium	420 ± 38	A01 B02 D01 E02
Phosphorus	171 ± 11	B02 B03 F01 F02
Chlorine	31 ± 21	D01 D04 K02
Zinc	18.6 ± 2.2	A01 A03 B02 B03 D01 D02 D03 E01 H01
Iron	14.8 ± 1.8	A01 B02 B03 D02 D03 E01
Strontium	4.62 ± 0.56	A01 B02 B03 D03 E01
Boron	2.8 ± 1.2	B02 B03 D04
Manganese	2.55 ± 0.29	A01 A05 B02 B03 D01 D03 E01 E02
Copper	2.47 ± 0.40	A01 A05 A06 B02 C03 C06 D01 D03 E01 H01
Barium	2.40 ± 0.52	B02 B03 C03 D01
Bromine	2.3 ± 0.5	D01 E01
Aluminum	1.01 ± 0.55	A05 A06 D01
Rubidium	0.5 ± 0.3	D01 D02 E01
Molybdenum	0.252 ± 0.039	B02 C03 C06 C07 D01 D03
Nickel	0.158 ± 0.054	A05 A16 C03 H01
Lead	0.140 ± 0.034	A04 A05 A16 C01 C03 H01
Chromium	0.101 ± 0.087	A06 B02 C05 D02 D03
Selenium	0.045 ± 0.008	C01 C04 D02 D03 G01
Iodine	0.026 ± 0.006	D03 D05 D06 F01
Cadmium	0.012 ± 0.005	A05 A16 C03 D03 H01
Cobalt	0.006 ± 0.006	A16 D01 D02 D03 H01
Vanadium	0.005 ± 0.002	B02 D01 D03
Mercury	0.003 ± 0.001	A09 A10 A15 D03
Arsenic	0.002 ± 0.002	A11 D03

[a] Prepared and Characterized by Agriculture and Agri-Food Canada. Distributed by the National Institute of Standards and Technology. From Ihnat (1993e).

TABLE 4   Example of elemental concentration values from the report of investigation for reference material corn bran (NIST RM 8433)^a (continued)

^{b} Best estimate values, weight percent or mg/kg (ppm), are based on the dry material, dried according to instructions in the report and are equally weighted means of results for generally at least two, but typically several different analytical methods applied by analysts in different laboratories. Uncertainties are estimates expressed either as a 95% confidence interval or occasionally (Al, B, Hg) as an interval based on the range of accepted results for a single future determination based on a sample weight of at least 0.5 g. These uncertainties, based on among-method, among-laboratory, among-unit, and within-unit estimates of variances, include measures of analytical method and laboratory imprecisions and biases and material inhomogeneity.

^{c} Analytical method codes and descriptions are provided in Table 5.

TABLE 5   Analytical methods used to determine best estimate and informational concentration values in reference material NIST RM 8433 Corn Bran^a

Code^b	Method name
A01	Acid digestion flame atomic absorption spectrometry.
A03	Dry ashing flame atomic absorption spectrometry.
A04	Acid digestion electrothermal atomic absorption spectrometry
A05	Closed vessel acid digestion electrothermal atomic absorption spectrometry.
A06	Dry ashing electrothermal atomic absorption spectrometry.
A09	Acid digestion cold vapor atomic absorption spectrometry.
A10	Closed vessel acid digestion cold vapor atomic absorption spectrometry with preconcentration.
A11	Closed vessel acid digestion hydride generation atomic absorption spectrometry with preconcentration.
A15	Acid digestion cold vapor atomic absorption spectrometry with preconcentration.
A16	Acid digestion coprecipitation electrothermal atomic absorption spectrometry.
B01	Acid digestion flame atomic emission spectrometry.
B02	Acid digestion inductively coupled plasma atomic emission spectrometry.
B03	Closed vessel acid digestion inductively coupled plasma atomic emission spectrometry.
C01	Acid digestion isotope dilution mass spectrometry.
C03	Closed vessel acid digestion isotope dilution inductively coupled plasma mass spectrometry.

TABLE 5 Analytical methods used to determine best estimate and informational concentration values in reference material NIST RM 8433 Corn Bran[a] (continued)

Code[b]	Method name
C04	Acid digestion dry ashing hydride generation isotope dilution inductively coupled plasma mass spectrometry.
C05	Dry ashing acid digestion isotope dilution mass spectrometry.
C06	Acid digestion isotope dilution inductively coupled plasma mass spectrometry.
C07	Dry ashing inductively coupled plasma mass spectrometry.
D01	Instrumental neutron activation analysis.
D02	Instrumental neutron activation analysis with acid digestion.
D03	Neutron activation analysis with radiochemical separation.
D04	Neutron capture prompt gamma activation analysis.
D05	Epithermal instrumental neutron activation analysis.
D06	Preconcentration neutron activation analysis.
E01	Particle induced X-ray emission spectrometry.
E02	X-ray fluorescence.
F01	Acid digestion light absorption spectrometry.
F02	Dry ashing light absorption spectrometry.
G01	Acid digestion fluorometry.
H01	Closed vessel acid digestion anodic stripping voltammetry.
I01	Kjeldahl method for nitrogen-volumetry.
I02	Kjeldahl method for nitrogen-light absorption spectrometry.
J01	Combustion elemental analysis-thermal conductivity.
J02	Combustion elemental analysis with chromatographic separation-thermal conductivity.
J03	Combustion elemental analysis-infrared spectrometry.
K02	Dry ashing volumetry.

[a] From Ihnat (1993e).
[b] Letter codes refer to classes of similar methods; number codes refer to specific variants.

ACKNOWLEDGMENT

Contribution No. 96-19 from Centre for Land and Biological Resources Research, Agriculture and Agri-Food Canada, Ottawa, Ontario, Canada.

REFERENCES

Bowen, H.J.M. 1985. Kale as a reference material, pp. 3–18. In: W.R. Wolf (Ed.), *Biological Reference Materials: Availability, Uses and Need for Validation of Nutrient Measurement.* John Wiley & Sons, New York, NY.

Bowman, W.S. 1994. *Catalogue of Certified Reference Materials, CCRMP 94-1E.* Canadian Certified Reference Materials Project, Natural Resources Canada, Ottawa, Canada.

Cantillo, A.Y. 1995. Standard and Reference Materials for Environmental Science, Parts 1 and 2. NOAA Technical Memorandum NOS ORCA 94 (Part 1 and Part 2), National Oceanic and Atmospheric Administration, Silver Spring, MD.

Cortes Toro, E., R.M. Parr, and S.A. Clements. 1990. Biological and Environmental Reference Materials for Trace Elements, Nuclides and Organic Microcontaminants, Report IAEA/RL/128 (Rev. 1). International Atomic Energy Agency, Vienna, Austria.

European Commission. 1996. *BCR Reference Materials* (Catalog). Institute for Reference Materials and Measurements (IRMM), Geel, Belgium.

Ihnat, M. 1986, 1993. Report of Analysis, Reference Material 8412 Corn (*Zea mays*) Stalk. National Institute of Standards and Technology, Gaithersburg, MD.

Ihnat, M. 1986, 1993. Report of Analysis, Reference Material 8413 Corn (*Zea mays*) Kernel. National Institute of Standards and Technology, Gaithersburg, MD.

Ihnat, M. 1988. Biological and Related Reference Materials for Determination of Elements, pp. 739–760. In: H.A. McKenzie and L.E. Smythe (Eds.), *Quantitative Trace Analysis of Biological Materials.* Elsevier Science, Amsterdam, The Netherlands.

Ihnat, M. 1993a. Reference Materials for Data Quality, pp. 247–262. In: M.R. Carter (Ed.), *Soil Sampling and Methods of Analysis.* Lewis Publishers, Boca Raton, FL.

Ihnat, M. 1993b. Report of Investigation, Reference Material 8416 Microcrystalline Cellulose. National Institute of Standards and Technology, Gaithersburg, MD; also: Report of Investigation, Reference Material MC 189 (NIST RM 8416) Microcrystalline Cellulose. Centre for Land and Biological Resources Research, Agriculture Canada, Ottawa, Ontario, Canada.

Ihnat, M. 1993c. Report of Investigation, Reference Material 8418 Wheat Gluten. National Institute of Standards and Technology, Gaithersburg, MD; also: Report of Investigation, Reference Material WG 184 (NIST RM 8418) Wheat Gluten. Centre for Land and Biological Resources Research, Agriculture Canada, Ottawa, Ontario, Canada.

Ihnat, M. 1993d. Report of Investigation, Reference Material 8432 Corn Starch. National Institute of Standards and Technology, Gaithersburg, MD; also: Report of Investigation, Reference Material CS 162 (NIST RM 8432) Corn Starch. Centre for Land and Biological Resources Research, Agriculture Canada, Ottawa, Ontario, Canada.

Ihnat, M. 1993e. Report of Investigation, Reference Material 8433 Corn Bran. National Institute of Standards and Technology, Gaithersburg, MD; also: Report of Investigation, Reference Material CB 186 (NIST RM 8433) Corn Bran. Centre for Land and Biological Resources Research, Agriculture Canada, Ottawa, Ontario, Canada.

Ihnat, M. 1993f. Report of Investigation, Reference Material 8436 Durum Wheat Flour. National Institute of Standards and Technology, Gaithersburg, MD; also: Report of Investigation, Reference Material DWF 187 (NIST RM 8436) Durum Wheat Flour. Centre for Land and Biological Resources Research, Agriculture Canada, Ottawa, Ontario, Canada.

Ihnat, M. 1993g. Report of Investigation, Reference Material 8437 Hard Red Spring Wheat Flour. National Institute of Standards and Technology, Gaithersburg, MD; also: Report of Investigation, Reference Material HRSWF 165 (NIST RM 8437) Hard Red Spring Wheat Flour. Centre for Land and Biological Resources Research, Agriculture Canada, Ottawa, Ontario, Canada.

Ihnat, M. 1993h. Report of Investigation, Reference Material 8438 Soft Winter Wheat Flour. National Institute of Standards and Technology, Gaithersburg, MD; also: Report of Investigation, Reference Material SWWF 166 (NIST RM 8438) Soft Wheat Flour. Centre for Land and Biological Resources Research, Agriculture Canada, Ottawa, Ontario, Canada.

Ihnat, M. 1994. Development of a New Series of Agricultural/Food Reference Materials for Analytical Quality Control of Elemental Determinations. *J. AOAC Int.* 77: 1605–1627.

Ihnat, M. 1997. Reference Materials for Data Quality Control, pp. 209–220. In: Y.P. Kalra (Ed.), *Handbook of Reference Methods for Plant Analysis*. St. Lucie Press, Boca Raton, FL.

Instytut Chemii i Techniki Jadrowej. Undated. Polish Certified Reference Material for Multielement Trace Analysis, Oriental Tobacco Leaves (CTA-OTL-1). Institute of Nuclear Chemistry and Technology (ICHTJ), Department of Analytical Chemistry, Warsaw, Poland.

International Atomic Energy Agency. 1982. Information Sheet for Reference Material V-8, Rye Flour. International Atomic Energy Agency, Vienna, Austria.

International Atomic Energy Agency. 1983. Information Sheet for Reference Material V-9, Cotton Cellulose. International Atomic Energy Agency, Vienna, Austria.

International Atomic Energy Agency. 1985. Information Sheet for Reference Material V-10, Hay Powder. International Atomic Energy Agency, Vienna, Austria.

International Atomic Energy Agency. 1994. Analytical Quality Control Services, Intercomparison Runs, Reference Materials 1994–1995. International Atomic Energy Agency, Vienna, Austria.

International Atomic Energy Agency. 1995. Survey of Reference Materials, Volume 1: Biological and Environmental Reference Materials for Trace Elements, Nuclides and Microcontaminants. IAEA-TECDOC-854, International Atomic Energy Agency, Vienna, Austria.

Muramatsu, Y. and R.M. Parr. 1985. Survey of Currently Available Reference Materials for Use in Connection with the Determination of Trace Elements in Biological and Environmental Materials, IAEA/RL/128. International Atomic Energy Agency, Vienna, Austria.

National Bureau of Standards. 1976a. Certificate of Analysis, Standard Reference Material 1570, Trace Elements in Spinach. National Bureau of Standards, Washington, D.C.

National Bureau of Standards. 1976b. Certificate of Analysis, Standard Reference Material 1573, Tomato Leaves. National Bureau of Standards, Washington, D.C.

National Bureau of Standards. 1976c. Certificate of Analysis, Standard Reference Material 1575, Pine Needles. National Bureau of Standards, Washington, D.C.

National Bureau of Standards. 1978a. Certificate of Analysis, Standard Reference Material 1567, Wheat Flour. National Bureau of Standards, Washington, D.C.

National Bureau of Standards. 1978b. Certificate of Analysis, Standard Reference Material 1568, Rice Flour. National Bureau of Standards, Washington, D.C.

National Bureau of Standards. 1982. Certificate of Analysis, Standard Reference Material 1572, Citrus Leaves. National Bureau of Standards, Washington, D.C.

National Institute of Environmental Studies (Japan). 1980. NIES Certified Reference Material CRM-1, Pepperbush. National Institute of Environmental Studies, Tsukuba, Japan.

National Institute of Environmental Studies (Japan). 1982. NIES Certified Reference Material CRM-3, Chlorella. National Institute of Environmental Studies, Tsukuba, Japan.

National Institute of Environmental Studies (Japan). 1986. NIES Certified Reference Material CRM-7, Tea Leaves. National Institute of Environmental Studies, Tsukuba, Japan.

National Institute of Environmental Studies (Japan). 1987. NIES Certified Reference Material CRM-10a, 10b, 10c, Rice Flour-Unpolished. National Institute of Environmental Studies, Tsukuba, Japan.

National Institute of Standards and Technology. 1988. Certificate of Analysis, Standard Reference Material 1567a, Wheat Flour. National Institute of Standards and Technology, Gaithersburg, MD.

Trahey, N.M. (Ed.). 1995. NIST Standard Reference Materials Catalog 1995–96, NIST Special Publication 260. Standard Reference Materials Program, National Institute of Standards and Technology, Gaithersburg, MD.

REFERENCE TEXTS ON PLANT ANALYSIS

Adriano, D.C. 1986. *Trace Elements in the Terrestrial Environment.* Springer-Verlag, New York.

Bennett, W.F. (Ed.). 1993. *Nutrient Deficiencies and Toxicities in Crop Plants.* APS Press, The American Phytopathological Society, St. Paul, MN.

Bergmann, W. 1992. *Nutritional Disorders of Plants: Development, Visual and Analytical Diagnosis.* Gustav Pischer Verlag, Jena, Germany.

Beverly, R.B. 1991. *A Practical Guide to the Diagnosis and Recommendation Integrated System (DRIS).* Micro-Macro Publishing, Athens, GA.

Black, C.A. 1993. *Soil Fertility Control and Evaluation.* Lewis Publishers, Boca Raton, FL.

Brown, P.H., R.M. Welsh, and E.E. Cary. 1987. Nickel: A micronutrient essential for higher plants. *Plant Physiol.* 85:801–803.

Chapman, H.D. (Ed.). 1966. *Diagnostic Criteria for Plants and Soils.* University of California, Division of Agriculture, Berkeley, CA.

Chapman, H.D. and P.F. Pratt. 1961. *Methods of Analysis for Soils, Plants, and Waters.* Priced Publication 4034. University of California-Berkeley, Division of Agriculture Sciences, Berkeley, CA.

Childers, N.F. 1966. *Nutrition of Fruit Crops, Temperate, Sub-tropical, Tropical.* Horticultural Publications, Rutgers—The State University, New Brunswick, NJ.

Gavlak, R.G., D.A. Horneck, and R.O. Miller (Eds.). 1994. *Plant, Soil, and Water Reference Methods for the Western Region.* Western Regional Extension Publication 125, University of California, Davis, CA.

Glass, A.D.M. 1989. *Plant Nutrition: An Introduction to Current Concepts.* Jones and Bartlett Publishers, Boston, MA.

Goodall, D.W. and P.G. Gregory. 1947. *Chemical Composition of Plants as an Index of their Nutritional Status. Imperial Bureau Horticultural Plantation Crops.* Technical Communications No. 17. Ministry of Agriculture, London, England.

Grundon, N.J. 1987. *Hungry Crops: A Guide to Nutrient Deficiencies in Field Crops.* Queensland Department of Primary Industries, Brisbane, Australia.

Halliday, D.J. and M.E. Trenkel (Eds.). 1992. *IFA World Fertilizer Use Manual.* International Fertilizer Industry Association, Paris, France.

Hardy, G.W. (Ed.). 1967. *Soil Testing and Plant Analysis.* Plant Analysis, Part II. Special Publication No. 2. Soil Science Society of America, Madison, WI.

Helrick, K. (Ed.) 1995. *Official Methods of Analysis by the Association of Official Analytical Chemists: Chemicals, Contaminants, Drugs.* Volume 1. Association of Official Analytical Chemists, Arlington, VA.

Johnson, C.M. and A. Ulrich. 1959. *Analytical Methods for Use in Plant Analysis.* Bulletin 766. University of California Agricultural Experiment Station, Berkeley, CA.

Jones, Jr., J.B. 1994a. *Plant Nutrition Manual.* Micro-Macro Publishing, Athens, GA.

Jones, Jr., J.B. 1994b. Plant Analysis (VHS video). St. Lucie Press, Boca Raton, FL.

Jones, Jr., J.B. 1994c. Tissue Testing (VHS video). St. Lucie Press, Boca Raton, FL.

Jones, Jr., J.B., B. Wolf, and H.A. Mills. 1991. *Plant Analysis Handbook: A Practical Sampling, Preparation, Analysis, and Interpretation Guide.* Micro-Macro Publishing, Athens, GA.

Kabata-Pendias, A. and H. Pendias. 1995. *Trace Elements in Soils and Plants.* Second edition. CRC Press, Boca Raton, FL.

Kitchens, H.B. (Ed.) 1948. *Diagnostic Techniques for Soils and Crops.* The American Potash Institute, Washington, D.C.

Marschner, H. 1986. *Mineral Nutrition of Higher Plants.* Academic Press, New York.

Martin-Prével, J. Gagnard, and P. Gauier. 1987. *Plant Analysis: As a Guide to the Nutrient Requirements of Temperate and Tropical Crops.* Lavoisier Publishing, New York.

Mengel, K. and E.A. Kirkby. 1982. *Principles of Plant Nutrition.* Third edition. International Potash Institute, Bern, Switzerland.

Mills, H.A. and J.B. Jones, Jr. 1996. *Plant Analysis Handbook II.* Micro-Macro Publishing, Athens, GA.

Mitchell, R.L. 1964. *The Spectrochemical Analysis of Soils, Plants, and Related Materials.* Technical Communications No. 44A. Commonwealth Agricultural Bureax, Farnham Royal, Bucks, England.

Mortvedt, J.J., P.M. Giordano, and W.L. Lindsay (Eds.). 1972. *Micronutrients in Agriculture.* Soil Science Society of America, Madison, WI.

Mortvedt, J.J., F.R. Cox, L.M. Shuman, and R.M. Welch (Eds.). 1991. *Micronutrients in Agriculture.* Second edition. SSSA Book Series 4. Soil Science Society of America, Madison, WI.

Pais, I. and J.B. Jones, Jr. 1996. *Handbook on Trace Elements in the Environment.* St. Lucie Press, Boca Raton, FL.

Piper, C.S. 1942. *Soil and Plant Analysis.* Hassell Press, Adelaide, Australia.

Plank, C.O. 1992. *Plant Analysis Reference Procedures for the Southern Region of the United States.* Southern Cooperative Series Bulletin 368. University of Georgia, Athens, GA.

Scaife, A. and M. Turner. 1984. *Diagnosis of Mineral Disorders in Plants.* Volume 2, Vegetables. Chemical Publishing, New York.

Sprague, H B. (Ed.). 1964. *Hunger Signs in Crops.* Third edition. David McKay Company, New York.

Walsh, L.M. and J.D. Beaton (Eds.). 1973. *Soil Testing and Plant Analysis.* Revised edition. Soil Science Society of America, Madison, WI.

Westerman, R.L. (Ed.). 1993. *Soil Testing and Plant Analysis.* SSSA Book Series 3. Soil Science Society of America, Madison, WI.

INDEX